Introduction to Biology

11th Hour

Introduction to Biology

David L. Wilson

Department of Biology
University of Miami
Coral Gables, Florida

Blackwell
Science

© 2000 by Blackwell Science Inc.
a Blackwell Publishing company

BLACKWELL PUBLISHING
350 Main Street, Malden, MA 02148-5020, USA
9600 Garsington Road, Oxford OX4 2DQ, UK
550 Swanston Street, Carlton, Victoria 3053, Australia

First published 2000

3   2006

Library of Congress Cataloging-in-Publication Data

Wilson, David L. (David Louis), 1943–
    Introduction to biology / David L. Wilson.
        p.cm.— (11th hour)
    ISBN 0-632-04416-0
    1. Biology. I. Title. II. Series: 11th hour (Malden, Mass.)
QH308.2.W524   2000
570–dc21   99-20229
                                                 CIP

ISBN-13: 978-0-632-04416-0

A catalogue record for this title is available from the British Library.

Set by Best-set Typesetter Ltd, Hong Kong

For further information on
Blackwell Publishing, visit our website:
www.blackwellpublishing.com

CONTENTS

11TH HOUR GUIDE TO SUCCESS

The 11th Hour Series is designed to be used when the textbook doesn't make sense, the course content is tough, or when you just want a better grade in the course. It can be used from the beginning to the end of the course for best results or when cramming for exams. Both professors teaching the course and students who have taken it have reviewed this material to make sure it does what *you* need it to do. The material flows so that the process keeps your mind actively learning. The idea is to cut through the fluff, get to what you need to know, and then help you understand it.

Essential Background. We tell you what information you already need to know to comprehend the topic. You can then review or apply the appropriate concepts to conquer the new material.

Key Points. We highlight the key points of each topic, phrasing them as questions to engage active learning. A brief explanation of the topic follows the points.

Topic Tests. We immediately follow each topic with a brief test so that the topic is reinforced. This helps you prepare for the real thing.

Answers. Answers come right after the tests; but, we take it a step farther (that reinforcement thing again), we explain the answers.

Clinical Correlation or Application. It helps immeasurably to understand academic topics when they are presented in a clinical situation or an everyday, real-world example. We provide one in every chapter.

Demonstration Problem. Some science topics involve a lot of problem solving. Where it's helpful, we demonstrate a typical problem with step-by-step explanation.

Chapter Test. For more reinforcement, there is a test at the end of every chapter that covers all of the topics. The questions are essay, multiple choice, short answer, and true/false to give you plenty of practice and a chance to reinforce the material the way you find easiest. Answers are provided after the test.

Check Your Performance. After the chapter test we provide a performance check to help you spot your weak areas. You will then know if there is something you should look at once more.

Sample Midterms and Final Exams. Practice makes perfect so we give you plenty of opportunity to practice acing those tests.

The Web. Whenever you see this symbol 🖥 the author has put something on the Web page that relates to that content. It could be a caution or a hint, an illustration or simply more explanation. You can access the appropriate page through *http://www.blackwellscience.com*. Then click on the title of this book.

The whole flow of this review guide is designed to keep you actively engaged in understanding the material. You'll get what you need fast, and you will reinforce it painlessly. Unfortunately, we can't take the exams for you!

PREFACE

Looking for that edge? Ready to learn biology, and master some of its more difficult concepts? Ready to work to improve your grade? If so, this book could be for you. The book reviews the more challenging material in a college-level, introductory course in biology. The book is intended as a supplement for a standard textbook in biology, or for students who wish to review such material.

Students who would benefit from this book might be enrolled in a major's or non-major's introductory biology course or could be taking advanced-placement biology. Others who would benefit might be studying for an exam based on such a course, such as the Medical College Admissions Test (MCAT) or an advanced-placement biology exam.

This book does not cover every topic that might be included in an introductory biology course. It dwells on the more difficult concepts—those that usually give students the most difficulty. I have tried to distill the core material to its essence, and provide ample opportunity to test your mastery of that material. There are few figures in the book because you have such figures in your text. If you are visually oriented, you probably will find it helpful to review such figures as you read this book. An asterisk (*) next to a problem or question indicates that it is especially difficult or challenging. Good luck and enjoy it.

Wherever you see () in the text, there is additional material on the Internet related to this topic. The Web site is located at http://www.blackwellscience.com. The site also contains many additional practice questions.

I thank Nancy Hill-Whilton, Irene Herlihy, and Jill Connor at Blackwell Science for their help and suggestions. It has been a pleasure working with them to bring one of Nancy Hill-Whilton's modules to life. I also thank the following reviewers for their numerous helpful suggestions for improving the presentation and coverage of this study guide: Sylvester Allred, Northern Arizona University; Wendy Bramlett, Sweet Briar College; Ann Lumsden, Florida State University; Holly Miner, University of South Dakota; and David Starret, Southeast Missouri State University. This book is dedicated to Mom, Dad, Peggy, and Mariah.

David L. Wilson, Ph.D.
Professor of Biology
University of Miami

UNIT I:
FROM ATOMS TO LIVING CELLS

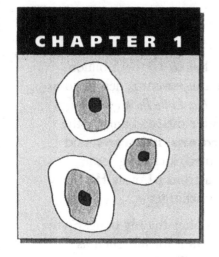

CHAPTER 1

Introduction to Biology

The understanding of life that has come in the past 150 years of modern biological research is a towering achievement, a central gift of the scientific enterprise to humanity. In this book I share that gift with you. Our tour of life begins in this chapter with several overviews. One considers how the material, the "stuff" of life came to be. A second introduces the major, established theories about the nature of life. We continue to explore such theories, and embellish them with details, throughout the rest of this book. Finally, the scientific method is reviewed. While most of you already have some knowledge of scientific method, there may be one or two more sophisticated twists that need to be mastered.

In understanding life it will be important to develop your ability to think on a number of levels. For instance, in terms of size, the average human height is somewhat less than 2 meters. Typical human cells are so small that it would take 40,000 of them, placed end-to-end, to span a meter. Typical bacterial cells are so small that it would take about 10 to 25 of them to cover the length of one human cell. Thus, bacterial cells can be as short as one **micron** (10^{-6} meters), compared to the 25-micron length of a typical human cell. The atoms that make up these cells are so small that thousands of them would be required just to form a line as long as a bacterial cell.

In terms of time, one-tenth of a second is fast for humans. That is our fastest reaction time, and human speed records are only kept to the nearest hundredth of a second. On a cellular level, a signal is passed from one nerve cell to another in about one-thousandth of a second. On a molecular level, events are faster still. Molecules collide with each other millions of times each second (one-millionth of a second is called a **microsecond**). We will see that life makes good use of the rapid rate of such collisions. Such a high rate of collisions can turn chance into near certainty.

ESSENTIAL BACKGROUND

- **Scientific notation**
- **Metric system of units**

TOPIC 1: A BRIEF HISTORY OF LIFE AND THE UNIVERSE

KEY POINTS

✓ *What do we know about the history of our universe?*

✓ *When did protons, neutrons, and electrons first form?*

✓ *Where were elements such as carbon, nitrogen, and oxygen formed?*

✓ *How old is life, and how long have humans been on earth?*

 We grow; we reproduce; we think. How did it all begin? Our universe began 10 to 15 billion years ago (1.5×10^{10} years ago; this is **scientific notation**). The initiating event is described as the **big bang**. For some fun reading on the subject consider Hogan's *The Little Book of the Big Bang* or Hawking's *A Brief History of Time*. During the first 3 minutes after the big bang, as the universe began to expand, energy density reduced enough to allow **protons**, **neutrons**, and **electrons** to form. Some of the very lightest of the elements also formed, including some helium and lithium. Protons are **hydrogen ions**, and so can be indicated as p^+ or as H^+. They are positively charged. Electrons are negatively charged; neutrons are uncharged.

The formation of stars and galaxies began during the first billion years after the big bang. Some continue to form, even today. Other than hydrogen, the elements important for life, such as carbon, nitrogen, and oxygen, were formed in stars by nuclear reactions. These elements then distributed across space as the early stars exploded into supernovas and spread the atoms necessary for life. Since many of the elements necessary for life were generated in stars, we are "stardust." Gravity caused the coming together of such stardust to form the earth. But we also have as part of us hydrogen atoms whose history may date all the way back to minutes after the big bang.

Our solar system, including earth, formed about 4.5 billion years ago. The first life on earth formed about 3.5 to 4 billion years ago. Complex, photosynthesizing organisms were present by 3.5 billion years ago.

Members of the genus *Homo* (human-like ancestors of ours) first walked the earth about 2 to 4 million years ago. (Please notice the thousand-fold shift here, from billions to millions.) Our species, *Homo sapiens*, first evolved about 200,000 years ago. The latest genetic evidence suggests that *Homo sapiens* emerged out of Africa about 100,000 years ago, spreading to Asia and Europe, and then to other continents. We all are related—there probably was but a single origin for all humans.

Topic Test 1: History of Life and the Universe

True/False

1. Most carbon atoms were formed during the big bang.

2. Humans are about as old as the earth is.

3. There are a million microns in 1 meter.

Multiple Choice

4. About how many years ago did life originate on earth?
   a. 3.5 to 4 thousand
   b. 3.5 to 4 million
   c. 3.5 to 4 billion
   d. 3.5 to 4 trillion
   e. None of the above

5. Consider the first stars that formed after the big bang. Could there have been planets with life on them around those original stars?
   a. Yes
   b. Only after millions of years had passed, so that life would have time to form
   c. No, because elements such as carbon, nitrogen, and oxygen were only first formed in the first stars
   d. No, because we are certain that the only life ever to exist in the universe is on earth

Short Answer

6. How many bacteria, 1 micron on each side, will fit in a typical human cell (about 25 microns on each side)?

For more problems on this topic, see the web page.

Topic Test 1: Answers

1. **False.** Carbon atoms are formed in stars.

2. **False.** Earth was formed about 4.5 billion years ago. Human-like organisms have been present for only 2 to 4 million years.

3. **True.** Each micron is one-millionth of a meter.

4. **c.** Life arose about a half-billion years after earth was formed.

5. **c.** Only second- and third-generation stars could be surrounded by planets with the elements necessary for life as we know it.

6. Although it is not necessary to do so, we can make a simplifying assumption that both human cells and bacterial cells have the shape of cubes. Since the bacterial cells are 1/25th the length of the human one, as many as $25 \times 25 \times 25$ will fit inside. (Picture a large cube and start stuffing it with the little ones, 25 rows of 25 on the bottom layer, and 25 such layers); $25 \times 25 \times 25 = 15,625$. Since we are estimating, we should round off to about 16,000. We will later see that this consideration of bacterial cells fitting within human cells is not without significance, both in terms of mitochondria and chloroplasts and in terms of our immune systems.

TOPIC 2: MAJOR GENERALIZATIONS ABOUT LIFE

KEY POINTS

✓ *What is evolution?*

✓ *Do living organisms obey the same laws as inanimate objects?*

✓ *What properties are shared among living organisms?*

✓ *Are there units from which living things are built?*

✓ *How do new properties emerge as we go from atoms to molecules, to cells, to organs, to organisms?*

1. Living things evolve. Not all species appeared at the same time. Instead, species come and go as life on earth evolves. In later chapters, we will explore the mechanisms, such as **natural selection**, that underlie these evolutionary changes, as well as consider some of the evidence that points to evolution occurring on earth. **Evolution** has become a very solid theory, central to understanding biology.

2. Living things obey the laws of physics and chemistry. It was once thought that life required special forces or spirits that went beyond those found in inanimate objects like rocks. **Vitalism**, the belief that special forces or spirits are involved in living organisms, has not been found to be necessary. We have been able to describe and understand the activities of living organisms solely on the basis of the principles of physics and chemistry. Biology is based on, and the science of biology builds from, physical principles.

3. Living things share certain properties. Among these properties are movement, growth, reproduction, complex organization and metabolism, responsiveness, use of energy, adaptation, and variation based on heredity. During the last 150 years, there has been a revolution in our understanding of biology. The science of biology has moved from collecting facts and descriptions of living organisms to understanding and explaining the properties of living organisms on a cellular and molecular level. The gathering of facts and descriptions is still continuing, and such efforts are important, but today we also can explain most of the essential functions of life—growth, reproduction, movement, metabolism, and adaptation—on a cellular or molecular level, as we will see in this book.

4. All living things are composed of cells. Cells, themselves, can be quite complex and organized. Some living organisms consist of just a single cell; others contain many cells. Each human has trillions of cells, all having arisen from a single fertilized egg cell.

5. Genes specify proteins, which in turn specify metabolism and structure. Genes serve as a kind of blueprint for the organization and functioning of living things. A typical gene contains the information specifying the structure for a particular protein. Some proteins serve a structural role in the cell, while others govern metabolism, stimulating chemical reactions that can result in the construction or destruction of other molecules.

New properties emerge as we progress up the hierarchy found in living organisms, from atoms to molecules, to cells, to tissues, to organs, to organisms, and finally to communities and ecosystems. The new properties that are present at higher levels appear to be the result of the nature and organization of the components at lower levels, and can be understood by considering the nature of the underlying components and the structures that these components form. Thus, molecules can have new properties not possessed by the individual atoms they are composed of, but these properties appear to result from the nature of the atoms in the molecules and the way the atoms are bonded together. We will see a good example of this as we examine the properties of water in Chapter 2. Similarly, we will see that cells exhibit new properties because of the organization and activities of subcellular structures. Such subcellular structures can contain large numbers of molecules organized to carry out the activities of the cell. None of this analysis by looking at the parts prevents scientists from directly studying higher levels as well. Science often gains insight by taking a top-down view as well as a bottom-up view of systems containing complex levels of organization. As we progress through this book, we will examine the emergence of new properties and will attempt to understand such properties at higher levels, but often we will see that one can use the underlying structure and order, where these are known, to enhance our understanding of the new properties, and of how they emerge from lower levels in the hierarchy.

Topic Test 2: Major Generalizations About Life

True/False

1. Proteins can be structural components in cells or can stimulate chemical reactions.

2. Some living organisms have no cells.

3. Some living organisms have only one cell.

4. All species on earth were created at the same time.

Multiple Choice

5. Which of the following is NOT correct?
   a. All living things are composed of cells.
   b. Living organisms are governed by vitalistic forces that differ from the forces that control inanimate objects.
   c. Living organisms share certain properties like growth, metabolism, and reproduction.
   d. Living organisms have evolved.

6. Which of the following is NOT one of the characteristics used to distinguish living organisms from inanimate objects?
   a. Growth
   b. Use of energy
   c. Dependence on other living organisms
   d. Reproduction
   e. Variation based on heredity

Short Answer

7. Part of the initial hypothesis of vitalism was the idea that organic compounds (compounds containing carbon) could not be produced without the aid of a "vital force" supplied by living organisms. In the mid-1800s several chemists made simple organic compounds from ordinary chemicals. What implications did such experiments have for ideas about vitalism?

Topic Test 2: Answers

1. **True.** Proteins play both roles in cells, and therefore play a very important role in living organisms.

2. **False.** All living organisms are composed of cells.

3. **True.** Living organisms, such as bacteria, consist of one cell.

4. **False.** Living organisms have evolved. We earlier discussed the idea that humans evolved only more recently. All life on earth today is thought to have originated from a single living entity. Today we see many species going extinct, often as a result of human activity. Several other mass extinctions occurred during the history of life on earth, and the number of species increased after such extinctions, as new species evolved.

5. **b.** There is no evidence that vitalistic forces are necessary to explain life.

6. **c.** While most living organisms depend on other organisms, creating a web of life, this is not one of the shared characteristics that are used to distinguish living organisms from nonliving organisms. Some living organisms are quite capable of surviving without other organisms being present.

7. Such experiments showed that nothing beyond ordinary chemistry was needed to form the special molecules found in living organisms. These experiments were the beginning of the end for the then-popular vitalistic view. Chemists have gone on to build an entire discipline of organic chemistry, creating many of the molecules of life and even new molecules, unlike any made in nature that we are aware of.

TOPIC 3: SCIENTIFIC METHOD

KEY POINTS

✓ *What methods are used by scientists to study nature?*

✓ *What is the difference between a hypothesis and a theory?*

✓ *What is falsifiability?*

In most textbooks about science, knowledge is presented as dogma. There is not much questioning of the information. Instead, you are expected to accept what is written as correct. This dogmatic approach is not the one used to gain scientific knowledge, but is efficient and perhaps the easiest way for you to be presented with what has been learned through science. Typically, only in more advanced courses do you learn any detail about the weaknesses in our knowledge or the doubts that some might have about particular theories.

The process of actually doing science is quite different from that of learning science. You probably have seen the typical sequence of steps that often underlie the **scientific method**: Axioms and assumptions → observations → formulate hypothesis → make predictions → perform tests → confirm, refute, or modify hypothesis → make further predictions and tests, and so on.

Scientists never prove anything to be true, but they can know some things beyond reasonable doubt. The steps noted above indicate that the search for truth in science is an ongoing process. If a hypothesis is significant and remains intact after a large number of tests, scientists will begin to call it a **theory**. A theory is something that has strong support, although even theories can be wrong or need modification if new tests reveal a mistake or limitation. Thus, the meaning that scientists give to the term **theory** differs from that used in ordinary speech, where one might say that such and such is "just a theory." Scientists would instead say "just a hypothesis."

Notice the role of axioms and assumptions in the sequence. These are necessary just to make observations. One thing that sets science apart from other knowledge systems is that the axioms and assumptions can themselves be subject to testing. Science is not built from an unchanging foundation. An axiom or assumption in one experiment can become the hypothesis being tested by another experiment.

Formulating a hypothesis is not just a matter of guessing. Scientists think carefully about their observations and try to produce a hypothesis that is testable and that will survive such tests. Hypothesis development is a creative endeavor, as is the designing of tests for such hypotheses.

Another way to distinguish scientific hypotheses from nonscientific ones is **testability**. To be scientific, a hypothesis must be capable of being tested. One aspect of testability is **falsifiability**, which states that for something to be considered scientific, there must be a test that could show the hypothesis to be wrong. Of course, if the test actually does this, then the hypothesis is wrong. However, for some hypotheses it is not conceivable that a test could refute the hypothesis. Any such hypothesis is not considered to be scientific.

For example, consider the "theory" of dreams put forward by Sigmund Freud. Freud's dream "theory" appears to be able to explain any dream that anyone has had. However, this actually becomes a weakness when one realizes that there is no dream that the so-called theory says one cannot have. As a consequence, it would appear that Freud's dream theory cannot be refuted by any test, since it does not rule out any possible dream. Therefore, many scientists do not even consider it a scientific hypothesis. A good dream hypothesis of this sort would specify dreams that we could not have. Then the hypothesis could be tested by examining whether such disallowed dreams ever occurred. This would make the hypothesis falsifiable—if anyone had a dream prohibited by the hypothesis, the hypothesis would be wrong. In fact, the more dreams that the hypothesis would declare impossible, the more impressive the hypothesis would be, presuming that it was not refuted, of course. Creation "science" is another example of a hypothesis that is not scientific.

Topic Test 3: Scientific Method

True/False

1. A hypothesis is stronger than a theory.

2. Once a test of a hypothesis has confirmed the hypothesis, no further tests are necessary.

3. One can make observations without first making any assumptions.

Multiple Choice

4. Falsifiability is
   a. the ability of a good scientist to prove any hypothesis false.
   b. the ability to demonstrate that a hypothesis is not scientific if it is shown to be false.
   c. the ability for a hypothesis to possibly be shown to be false by a test.
   d. a helpful way to distinguish scientific and nonscientific hypotheses.
   e. Both c and d are correct.

Short Answer

5. Consider a set of cards. Make the assumption that each card has a letter on one side and a number on the other. The hypothesis to be tested is as follows: If there is an A on one side of the card, then there is a 3 on the other side. Which two of the following four cards offer the best tests of this hypothesis (which two would it be best to "turn over" to see what is on the other side)?

$$\boxed{A} \quad \boxed{B} \quad \boxed{3} \quad \boxed{4}$$

Topic Test 3: Answers

1. **False.** Only after a hypothesis has been well confirmed by experimental tests can it be called a theory.

2. **False.** A single confirmation of a hypothesis is but a start on the scientific method of establishing hypotheses as correct. A variety of tests, and repetitions of tests, must occur.

3. **False.** Any observation requires prior assumptions related to what one is observing.

4. **e.** Falsifiability is both the ability of a hypothesis to be really tested and a way to identify some nonscientific hypotheses.

5. The first card, "A," should be turned over because a number other than 3 on the other side would falsify the hypothesis, making this a good test of the hypothesis. The second card, "B," should not be turned over because it is not a test of the hypothesis. Even were there to be a "3" on the other side, it would not refute the hypothesis, since the hypothesis says nothing about what should be on the other side of a "B" card. The third card could serve to confirm the hypothesis, but it should not be turned over because there is no way that the hypothesis can be refuted by turning it over. Thus, it does not really present a good test of the hypothesis. The hypothesis does not say if there is a "3" on one side, there will be an "A" on the other. The hypothesis does not indicate what has to be on the other side of cards with a "3" on them. Be sure that you understand the meaning of "if . . . then" statements, since they are often used in science. The fourth card, "4," should be turned over because it could refute the hypothesis if there is an "A" on the other side. Thus, it is a good test of the hypothesis.

APPLICATION: SERENDIPITY IN SCIENTIFIC RESEARCH

Although scientists attempt to control experimental conditions carefully, things can go wrong. Usually, scientists will reject and discard results when something goes wrong. However, occasionally, mistakes or unavoidable variations can lead to new discoveries. Such was the case with Alexander Fleming, who was in the middle of an experiment when he came across a petri dish that had a contaminating mold growing amid bacteria. Bacteriologists of that day usually would reject such a contaminated plate and move on, but Fleming noticed that there was a clear area surrounding the mold where no bacteria were present. Fleming realized that a chemical produced by the mold might be killing the nearby bacteria. He isolated the substance and named it **penicillin**, which became the first antibiotic.

I had my own experience with serendipity when Hurricane Andrew hit south Florida in 1992. I was in the middle of an experiment studying a mutant in a nematode that allowed the nematode to live longer than normal. I did not know why the mutant was able to live longer, and was trying to figure out what was different about the mutant. For a few days after the hurricane, there was no air conditioning in the laboratory because the power was out. As a result, the temperature increased, and my normal nematodes died more rapidly than normal. However, the animals with the aging mutation lived about as long as they did when temperatures were lower. I realized that the mutation must have allowed the animals to survive the heat stress better. I told others about my observation and, within a couple of years, they confirmed that the mutant animals were generally more resistant to such stresses as heat. Without my bit of serendipity, it might have been much longer before we gained insight into what the mutant was doing. This is about the only good thing that resulted from Hurricane Andrew!

Chapter Test

True/False

1. A human contains trillions of cells.

2. Living organisms obey the same physical and chemical laws as inanimate objects.

3. Evolution took place on earth until humans appeared, but now has ceased.

Multiple Choice

4. Which of the following is correct?
   a. Some living things have no cells.
   b. Living organisms are vitalistic; that is, they are governed by special forces that do not follow physical and chemical laws.
   c. Genes specify the kinds of proteins a cell can make, and proteins help to determine the metabolism and structures in living organisms.
   d. A hypothesis is an observation that has been shown to be correct.
   e. There is no evidence to support evolution since it is historical and only happened once.

5. *Homo sapiens* have been on earth
   a. since the time that dinosaurs were on earth.
   b. as long as frogs have been on earth.
   c. for about 10,000 human generations.
   d. for 1 billion years.
   e. as long as the earth has been here.

6. Science
   a. proves theories.
   b. tests hypotheses.
   c. establishes truth.
   d. has laws that never change.
   e. does all of the above.

7. Freud's dream "theory" is thought not to be scientific by some because
   a. it is not falsifiable.
   b. it is too specific.
   c. it deals with sex, not science.
   d. Freud was not a scientist.
   e. it has been shown to be false.

Short Answer

8. What distinguishes living from nonliving things?

Chapter Test Answers

1. **T** 2. **T** 3. **F** 4. **c** 5. **c** 6. **b** 7. **a**

8. Living things possess sets of properties that are not possessed by nonliving things. Nonliving things can possess some of the properties (cars move), but do not possess the set (cars do not reproduce). Not every living thing possesses every property on the list (worker bees cannot reproduce), but they have enough of the properties to allow one to distinguish them as living. A list of such properties is given under topic 2, above.

Check Your Performance:

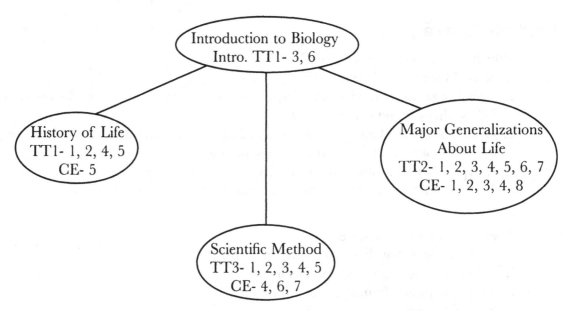

Key: TT = Topic Test; CE = Chapter Exam. Numbers indicate exam questions. Some questions are listed more than once if they refer to more than one topic.

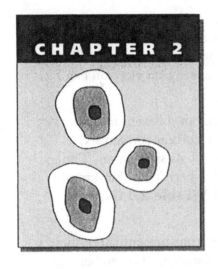

CHAPTER 2

Atoms, Bonds, Water, and Carbon

Chemistry and physics form the foundation for biological sciences. Atoms and molecules are the materials of life. In this chapter we come to understand atoms and how they combine to form molecules. We also examine the special properties of water and carbon that make them so important in living organisms.

ESSENTIAL BACKGROUND

- **Logarithms**

TOPIC 1: ATOMS

KEY POINTS

✓ *What are atoms composed of?*

✓ *How do elements differ from one another?*

✓ *What are isotopes? What is radioactivity?*

✓ *What is a half-life?*

Each **atom** is composed of a nucleus, containing one or more protons (p^+). Most atomic nuclei also contain neutrons (n). Atoms are surrounded by electrons (e^-), which form a kind of cloud around the small nucleus at the center of the atom. Protons are positively charged; electrons carry an equal but opposite, negative charge. **Ions** are charged atoms; that is, they carry an electrical charge because the number of protons does not equal the number of electrons. Usually, electrons have been lost or added to the atom to create the ion.

The number of protons in an atom or ion, the **atomic number**, identifies the kind of atom, or **element**. The most abundant elements in living organisms are **carbon (C)**, **hydrogen (H)**, **nitrogen (N)**, **oxygen (O)**, **phosphorus (P)**, and **sulfur (S)**. These comprise about 99% of the mass of living organisms and can be remembered by the letters CHNOPS (think "chin-ups"). In humans, **calcium (Ca)**, which is abundant in bone, is included to account for 99% of the mass of a human. A few other elements we discuss in this book are important in living organisms. Among these are the ions found in our body fluids, such as **sodium (Na^+)**, **chloride (Cl^-)**, and **potassium (K^+)**. **Iron (Fe)** plays an important role as part of a few different kinds of proteins, including hemoglobin, which transports oxygen in blood.

The electrons in an atom are in orbits around the atomic nucleus. The electron orbits are organized into shells and subshells. As an electron gains energy, it will orbit farther from the nucleus—the higher the energy level, the less tightly bound the electron is to the nucleus. Electrons lose energy as they drop back toward the nucleus. With enough energy, an electron can be ejected from the atom.

Isotopes are atoms with the same number of protons (same element), but different numbers of neutrons. For instance, carbon has six protons, but different isotopes of carbon contain six, seven, or eight neutrons. The atomic number for carbon is six because of its six protons. The carbon isotope with six protons plus six neutrons is symbolized by ^{12}C, where 12 is the sum of the number of protons and neutrons. This isotope is called **carbon 12**. **Table 2.1** lists some properties of CHNOPS.

Isotopes have different masses because of the different numbers of neutrons present. Also, they can have different levels of stability against **radioactive decay**. For instance, ^{12}C is stable and does not decay, but ^{14}C is radioactive and decays by releasing an electron from the nucleus. (In a simplified way, you can imagine that a neutron has split to give an electron, which flies out of the nucleus and leaves a proton behind. A neutrino also is released, which we can ignore for our purposes.) Thus, when ^{14}C decays, it leaves behind an atomic nucleus that contains seven protons. Such a nucleus is no longer carbon but has become nitrogen.

Radioactive isotopes are used by biologists as tools for a range of experiments, from determining the age of fossils to studying biochemical processes in living cells. Later we will see that radioactive isotopes were used to determine the nature of genetic material, and that stable isotopes were used to determine the way that the genetic material gets copied.

Radioactive decay of atoms occurs over time as (a negative) exponential. This means that in a group of identical radioactive isotopes of an element, the time for half of them to decay is a constant. This constant is called the **half-life** for the isotope. The half-life for ^{14}C is over 5,700 years. Thus, if we had 10 billion atoms of ^{14}C at a given time, then we would have about 5 billion left 5,700 years later.

Radioactive decay is a nuclear event, the kind of event that generates nuclear energy and atomic bombs. In contrast, living things depend mostly on chemical events, which are more concerned with the electron clouds around atomic nuclei. Chemical reactions are the changing or rearranging of electrons around groups of atomic nuclei, and not changes within the atomic nuclei themselves.

Table 2.1 Properties of Abundant Elements Found In Living Organisms

Element	Symbol	Protons[a]	(Protons + Neutrons)[b]	Approximate Atomic Weight	Bonds[c]
Hydrogen	H	1	1, 2, <u>3</u>	1	1
Carbon	C	6	12, 13, <u>14</u>	12	4
Nitrogen	N	7	14, 15	14	3
Oxygen	O	8	16, 17, 18	16	2
Phosphorus	P	15	31	31	5
Sulfur	S	16	32, 33, 34	32	2

[a] The number of protons also is the atomic number.
[b] These list the isotopes commonly found on earth. The radioactive isotopes are underlined. They are rare unless generated artificially.
[c] The number of covalent bonds formed (see next section).

The **atomic weight (atomic mass)** of an element is approximately equal to the sum of the number of neutrons plus the number of protons in the atoms of the element. Thus, ^{12}C would have an atomic weight of 12. It also happens that the atomic weight for an element, measured in grams, always contains 6.02×10^{23} atoms. This is **Avogadro's number**. For example, 12 grams of ^{12}C contains 6.02×10^{23} atoms of carbon. Because most elements in nature consist of more than one isotope, the atomic weight is an average, based on the fraction of each isotope present. Thus, carbon consists mostly of ^{12}C, but with a little ^{13}C and a bit of ^{14}C, so the actual atomic weight for carbon is 12.01.

Topic Test 1: Atoms

True/False

1. Radioactive decay of an atom is caused by an unstable array of electrons in the atom.

2. According to Table 2.1, some sulfur atoms contain 16 protons and 16 neutrons.

3. According to Table 2.1, ^{32}P is not an abundant isotope of phosphorus.

Multiple Choice

4. The isotopes of an element have different numbers of
   a. neutrons and the same number of protons.
   b. protons and the same number of neutrons.
   c. electrons and the same number of neutrons and protons.
   d. electrons and different numbers of protons.

5. One gram of hydrogen contains about as many atoms as
   a. 1 gram of carbon atoms.
   b. 1 gram of oxygen atoms.
   c. 12 grams of ^{12}C.
   d. 12 grams of ^{14}C.
   e. Both a and b are correct.

Short Answer

6. If the half-life of ^{14}C is about 5,700 years, how much will be left after two half-lives, due to radioactive decay?

Topic Test 1: Answers

1. **False.** Radioactive decay is caused by an unstable nucleus, which is not related to the electrons in the atom.

2. **True.** Since all sulfur atoms have 16 protons and since one of the isotopes for sulfur contains 32 protons and neutrons according to Table 2.1, then this isotope must contain 16 neutrons as well.

3. **True.** The isotope ^{32}P actually is radioactive, and artificially generated ^{32}P is sometimes used by scientists to tag or label certain biological molecules, such as nucleic acids. It decays rapidly, with a half-life of about 2 weeks, and so little remains naturally on earth.

4. **a.** Isotopes are defined as being the same element, but having different numbers of neutrons. The isotopes of an element will have the same number of protons since elements are defined by the number of protons they contain.

5. **c.** The atomic mass, in grams, contains Avogadro's number, 6.02×10^{23} atoms. Since the atomic mass of hydrogen is 1, there are 6.02×10^{23} atoms in 1 gram of hydrogen. Of the listed options, only 12 grams of ^{12}C contains the same number of atoms, since it also contains Avogadro's number of atoms.

6. In 5,700 years half of the atoms will remain. After one half-life, one can start again with half the initial amount and so on. In one more half-life, half again will be gone, leaving one-fourth. So, one-fourth of the atoms will remain after two half-lives, or 11,400 years. Notice that the radioactivity is not all gone after two half-lives. Each atom has a constant probability of decaying with time.

TOPIC 2: CHEMICAL BONDS

KEY POINTS

✓ *What are molecules? How is molecular weight determined?*

✓ *What are ionic and covalent bonds?*

✓ *What is a polar, covalent bond?*

Molecules are bonded groups of atoms. The **molecular weight (molecular mass**, in units of **daltons**) is the sum of the atomic weights of the atoms in the molecule. The molecular weight in grams always contains 6.02×10^{23} molecules. It is called 1 mole. A **mole** is just a convenient way of referring to Avogadro's number of molecules. For example, water, H_2O, which consists of two atoms of hydrogen bonded to one atom of oxygen, has a molecular weight of 18 (each hydrogen has an atomic weight of about 1 and oxygen has an atomic weight of 16—8 protons plus 8 neutrons—see Table 2.1; 2 plus 16 equals 18). Eighteen grams of water is 1 mole of water and contains Avogadro's number—6.02×10^{23}—of molecules.

As atomic number increases, there is a periodic repeating of the chemical properties of the elements. Thus, carbon (6 protons) has chemical properties that are similar to silicon (14 protons), and oxygen (8 protons) has properties similar to sulfur (16 electrons), as predicted from their positions in the same column of the periodic table. This periodicity, or repeating, of the chemical properties is what leads to the **periodic table** of the elements, and is at the foundation of the science of chemistry. The chemical bonding properties of atoms can be predicted by their electron configurations. The number of bonds made by each of the common elements making up living organisms is listed in Table 2.1. You should know these.

There are two principal kinds of strong bonds in living organisms. One is the **ionic bond**. Ionic bonds are formed by the mutual attraction of ions of opposite charge—one positive, the other negative. An example of an ionic bond is that in the salt sodium chloride (NaCl). Each molecule of NaCl contains one sodium ion (Na^+) and one chloride ion (Cl^-). Sodium atoms tend to lose an electron, resulting in a +1 electrical charge, while chloride atoms gain an electron and a −1 electrical charge. The formation of an ionic bond requires the gaining or losing of one or more electrons by atoms or groups of atoms forming the ionic bond.

When sodium chloride, which is table salt, is placed in water, it dissolves into separate Na^+ and Cl^- ions in the water solution. This occurs because of the special properties of water, which we will consider later.

The second major kind of bond is the **covalent bond**. Covalent bonds involve the sharing of electrons between atoms or groups of atoms. Perhaps the simplest example of a covalent bond is that between two atoms of hydrogen, which will form a bond that involves the sharing of a pair of electrons, one from each hydrogen atom. Usually such a covalent bond is indicated by a bar or line connecting the two atoms such as: H—H.

Covalent bonds involve the sharing of pairs of electrons, often a single pair, but sometimes two or three pairs. When two pairs of electrons are shared, a double bond is formed. Carbon dioxide (CO_2) consists of two double bonds—each oxygen is linked to a central carbon by two double bonds (O=C=O). The three atoms in carbon dioxide are in a line with one another (bond angle is 180 degrees). When three pairs are shared, it is a triple bond.

In some covalent bonds, the electrons are shared rather equally. This is the case when carbon atoms bond with other carbon atoms or with hydrogen atoms. However, in other covalent bonds, the electrons are not shared equally. These are called **polar covalent bonds**. Polar bonds often lead to partial charge separations—the unequally shared electrons will spend more time around one atomic nucleus than around the other nucleus. Oxygen atoms, as well as nitrogen and sulfur, are called **electronegative** because they tend to attract shared electrons more strongly than do most atoms. When linked to carbon or hydrogen, they can produce polar bonds. For example, such polar bonds exist in water molecules. The central oxygen atom will have a slight negative charge on it and the two hydrogen atoms that are attached to the oxygen will have a slight positive charge. The bond angle for water, unlike that for carbon dioxide, is not 180 degrees. Instead, the molecule is bent at an angle, leaving a charge separation between the oxygen side and the hydrogen side of the molecule. This plays an essential role in giving water, and life, its properties, as we will see.

Topic Test 2: Chemical Bonds

True/False

1. Ionic bonds involve the sharing of electrons.
2. When sharing of electrons is unequal, we say that we have a polar ionic bond.

Multiple Choice

3. A polar bond is
   a. any chemical bond found at the North or South Pole.
   b. produced when carbon bonds to hydrogen.
   c. a covalent bond with equal sharing of electron pairs.
   d. a bond with unequal sharing of electron pairs.
   e. an ionic bond.

4. Eighteen grams of H_2O contains how many molecules?
   a. 18
   b. 54

c. 3.02×10^{23}

d. Within a very small error, the same as the number of atoms in 12 grams of ^{12}C

e. More than in 5 grams of hydrogen molecules (H_2)

Short Answer

5. Use the information in Table 2.1 to determine how much a mole of $C_6H_{12}O_6$ weighs.

6. How many molecules are in half a mole?

Topic Test 2: Answers

1. **False.** Ionic bonds are between atoms that have gained or lost full electrons. Covalent bonds involve shared electrons.

2. **False.** Unequal sharing of electrons produces polar covalent bonds.

3. **d.** Unequal sharing produces a polar bond. Carbon bonds to hydrogen with a nonpolar bond.

4. **d.** Eighteen grams of water equals 1 mole of water. Twelve grams of ^{12}C equals 1 mole of carbon atoms. None of the others are 1 mole.

5. We can make use of the atomic weights of each of the three kinds of atoms, and multiply them by the number of atoms of each type within the molecule. This will give: C, 12×6; H, 1.0×12; O, 16×6. The sum of the three products is 180.16.

6. A mole contains 6.02×10^{23}, so half a mole contains 3.01×10^{23} molecules.

TOPIC 3: WATER: pH AND HYDROGEN BONDS

KEY POINTS

✓ *How do hydrogen bonds explain the special properties of water that are so important for living organisms?*

✓ *What is pH a measure of, and what scale is used?*

Properties of Water

Water is the most abundant molecule in almost all living organisms, making up more than 50% of the total weight of organisms. Water has a number of unusual properties. For its size, water has a very high boiling point. Most molecules with the mass of water have a boiling point below that of typical body temperatures. Obviously, life as we know it could not exist on earth if water were to have a similar boiling point.

Water also has a high heat content: It takes a lot of heat to warm water. If this were not true, our body temperatures would not be able to remain as constant as they do, and that also would cause us real problems.

Other properties of water are special: Water clings to itself. Water's cohesion leads to its properties of high surface tension and capillary action. The high surface tension allows water striders (insects) to skate across the surface of water without breaking through the surface. Its capillary action is used by plants to aid in the movements of dissolved substances.

Water has special properties as a solvent, as other polar or charged molecules usually dissolve readily (are soluble) in it, while nonpolar, uncharged substances such as oils remain insoluble in water (consider vinegar-and-oil salad dressing, and how the vinegar, or water, phase remains separate from the oil phase). The water-insoluble molecules are referred to as **hydrophobic** ("water avoiding") as opposed to **hydrophilic** ("water loving") polar molecules that dissolve easily in water. Hydrophobic molecules include nonpolar molecules with an abundance of hydrocarbon (hydrogen plus carbon) chains. We will see that such molecules and their insolubility in water play an important role in membranes, the outer boundaries of living cells.

Hydrogen Bonds

What produces the special properties of water that were just described? Water consists of polar, covalent molecules. Oxygen is electronegative—it strongly attracts electrons. The shared pair of electrons between oxygen and each attached hydrogen in an H—O—H molecule spends more time around the oxygen than around the hydrogen. Thus, each water molecule has a charge separation within it (remember that water molecules are not linear, but bent). In water, the charge separation allows the H_2O molecules to attract one another. The negatively charged oxygen atom in one molecule is attracted to the positively charged hydrogen atoms of nearby molecules. Weak bonds, called **hydrogen bonds**, are formed among these different water molecules. Hydrogen bonds have about one-tenth the strength of a typical covalent bond, such as that between hydrogen and oxygen in individual water molecules. We will see in later chapters that hydrogen bonds also play essential roles in the structure of such molecules as DNA and protein.

The numerous hydrogen bonds between neighboring water molecules give water its special properties: Water boils at a higher temperature because boiling entails individual water molecules escaping from the liquid into the vapor phase, and this requires the breaking of the hydrogen bonds. Similarly, the hydrogen bonds contribute to water's cohesiveness and high surface tension; the bonds must be broken if an insect is to break through the surface of water. The hydrogen bonds must be stretched when water is heated. That stretching takes energy, and so it takes extra energy to increase the temperature of water. The hydrogen bonds link neighboring water molecules, and energy is required to increase the temperature, which results in increased vibrations between water molecules (such vibrations stretch the bonds). Even the solubility properties of water are related to oxygen's electronegativity and the hydrogen bonds. Other polar substances can substitute for the hydrogen bonds as they solubilize and become part of a water-based solution, but nonpolar molecules could only intermingle with water molecules by disruption of some of the hydrogen bonds, and that requires energy. The lower energy state is to exclude the nonpolar substances, and that is what happens when oil and water do not mix.

We have explained the "emergent" properties of water.

pH

Another special property of water is the splitting of a small fraction of the molecules that occurs in water and water solutions. The molecules will split as follows:

$$H_2O \rightarrow H^+ + OH^-$$

(The H^+ actually immediately binds to another water molecule to form H_3O^+, the hydronium ion, but for simplicity we will ignore this complexity and refer to hydrogen ions.) Because of this splitting, water contains a small amount of hydrogen ions, or protons (H^+ is just a proton), and hydroxyl ions (OH^-). We have a special way of referring to the concentration of hydrogen ions in water; we use **pH** to do so.

$$pH = -\log[H^+],$$

where log stands for logarithm to the base 10, and [ ] stands for "concentration of." Thus, this equation states that the pH of a solution is the negative logarithm of the concentration of hydrogen ions in the solution. The concentration of such ions is expressed in moles per liter. An equivalent equation is:

$$[H^+] = 10^{-pH},$$

which can be read as follows: The concentration of hydrogen ions is equal to 10 raised to the negative pH power. So, pH gives us a shorthand way of talking about the concentration of hydrogen ions in water solutions. Pure water has **neutral pH**, which is pH 7. This means that the hydrogen ion concentration in pure water is 10^{-7} moles/liter. **Acidic** solutions have a lower pH (higher concentrations of hydrogen ions). **Basic** solutions have a higher pH (lower concentrations of hydrogen ions). Thus, acid means pH < 7, and the stronger the acid, the lower the pH; neutral means pH 7; base means pH > 7, and the stronger the base, the higher the pH.

In all water solutions, the following equation also holds true:

$$[H^+][OH^-] = 10^{-14} \text{ (moles/liter)}^2$$

This equation indicates that the product of the concentration of hydrogen ions times the concentration of hydroxyl ions is a constant. Another way of saying the same thing is **pH + pOH = 14**, where pOH is defined for hydroxyl ions the same way that pH is defined for hydrogen ions. This equation can be used to determine either hydrogen or hydroxyl concentration in a solution, when given the other. Thus, if pH is 7, then pOH must be 7 as well. At neutral pH, the concentration of hydrogen ions is equal to the concentration of hydroxyl ions. In acidic solutions, the concentration of hydrogen ions is greater than that of hydroxyl ions. In basic solutions, the opposite is true.

Buffers resist changes in pH in a solution. Buffers are substances that take up or release hydrogen ions or hydroxyl ions to offset the addition of the opposite (hydroxyl or hydrogen ions) into a solution. Buffers tend to keep pH more constant. For example, if you add an acid or a base to pure water, you will get a quick change in pH. However, if there is a buffer, such as bicarbonate ion (HCO_3^-), the pH might change very little, because with the addition of acid (H^+):

$$H^+ + HCO_3^- \rightarrow H_2CO_3 \rightarrow H_2O + CO_2$$

and with the addition of base (OH^-):

$$OH^- + HCO_3^- \rightarrow CO_3^{-2} + H_2O$$

In both cases the added acid or base is neutralized by the bicarbonate ion.

Blood, other body fluids, and cells contain buffers (bicarbonate ions, proteins, etc.) that keep a near-constant pH in the body, except in certain diseases and disorders. The near-constant pH is critically important to proper functioning of living organisms.

Topic Test 3: Water

True/False

1. If the pH of a solution is 5, then the solution is acidic.

2. If the pH of a solution is 5, then the concentration of hydrogen ions in the solution is 10^{-5} moles/liter.

3. If the pH of a solution is 5, then the hydroxyl ion concentration in the solution is 10^{-9} moles/liter.

4. Hydrogen bonds contribute to the high boiling point of water.

Multiple Choice

5. A solution has a concentration of hydroxyl ions of 10^{-3} moles/liter. What is the pH of the solution?
   a. 3
   b. 5
   c. 7
   d. 9
   e. 11

6. The pH of a solution is 3. The solution
   a. is a basic, or alkaline, solution.
   b. is more basic than a solution with a pH of 7.
   c. has 10 times the hydrogen ion concentration of a solution with a pH of 2.
   d. has a hydrogen ion concentration of 10^{-7} moles/liter.
   e. has a hydroxyl ion concentration of 10^{-11} moles/liter.

7. Which of the following is an example of a hydrogen bond?
   a. The bond between two hydrogen atoms
   b. The bond between a hydrogen and carbon atom in an oil molecule
   c. The bond between a hydrogen atom on one water molecule and the oxygen atom on another water molecule
   d. The bond between sodium and chloride atoms in salt
   e. All of the above

Short Answer

8. Why are nonpolar molecules not very soluble in water?

Topic Test 3: Answers

1. **True.** Any pH < 7 is acidic, as hydrogen ion concentration is greater than hydroxide ion concentration.

2. **True.** This is the definition of pH.

3. **True.** A pH of 5 means a hydrogen ion concentration of 10^{-5} moles/liter. But the product of hydrogen and hydroxyl ion concentrations must be 10^{-14} moles/liter, so

$$[OH^-] = 10^{-14}/10^{-5} = 10^{-9} \text{ moles/liter}$$

4. **True.** The hydrogen bonds between water molecules must be broken for water to change from the liquid to the gas phase, and that requires extra energy, which is supplied by a higher temperature of boiling.

5. **e.** The pH is 11. If $[OH^-] = 10^{-3}$ moles/liter, then

$$[H^+][OH^-] = 10^{-14},$$
$$[H^+]10^{-3} = 10^{-14}, \text{ or}$$
$$[H^+] = 10^{-11} \text{moles/liter, or pH} = 11.$$

6. **e.** The reasoning is the same as in question 5. For answer c, notice that at pH 3, the hydrogen ion concentration is actually 10 times less than at pH 2.

7. **c.** Only answer c lists a hydrogen bond. However, hydrogen bonds also can form between hydrogen and other electronegative ions, such as nitrogen and sulfur.

8. Nonpolar molecules are not very soluble in water because they would have to disrupt hydrogen bonds in order to mix with the water molecules. Since this would require energy, most of the nonpolar molecules will be excluded from the water phase.

TOPIC 4: CARBON: TETRAHEDRAL BONDS AND ISOMERS

KEY POINTS

✓ *How many covalent bonds do carbon atoms form with other atoms, and why is this number important for life?*

✓ *What is the shape of the bonds that a carbon atom forms with other atoms?*

✓ *What are isomers? What are the different kinds of isomers?*

Carbon atoms are critical to life as we know it. Carbon is found in most of the molecules that make up living organisms. What makes carbon so important is the fact that each carbon atom can bond with up to four other atoms. This allows a great variety of structures to be formed from a few different kinds of atoms, as we will see in Chapter 3. We can begin to describe a molecule by indicating its **molecular formula**. Molecular formulas are formulas like H_2O or CH_4. **Structural formulas** give more information, indicating the bonds between the individual atoms in a molecule. Some examples of structural formulas are given in **Figure 2.1**.

However, even structural formulas do not tell the whole story because of the nature of the bonds formed by carbon atoms. When a carbon atom has four single bonds, the bonds are in the shape of a **tetrahedron**. Tetrahedrons are three-dimensional structures with a triangle on each of its four sides. The four atoms or groups linked to a single carbon atom can be viewed as being at the four corners of the tetrahedron. **Figure 2.2a** shows such a structure, with a carbon atom in the middle of the tetrahedron, linked to four imaginary atoms or groups of atoms, W, X, Y, and Z. Notice that W, X, Y, and Z are not all in a plane. You can view Z as coming out of the paper toward you. Thus, if W, X, Y, and Z are four different atoms or groups of atoms, then two molecules are possible mirror images of each other, as shown in **Figure 2.2b**. These two molecules

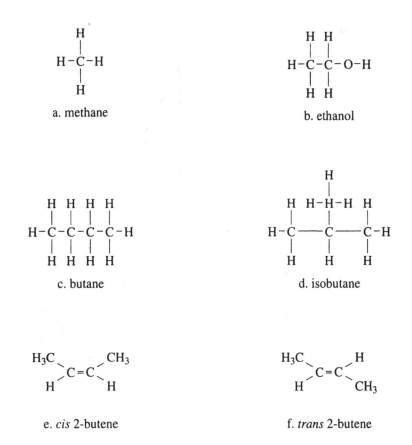

Figure 2.1 Structural formulas. Butane and isobutane (c and d) are structural isomers. They have the same chemical formula (C_4H_{10}). Parts e and f show geometric isomers. Notice the different arrangement of the methyl groups in the two molecules. The groups cannot rotate around the double bond. As expected of isomers, this pair also has the same chemical formula (C_4H_8).

are not alike, as can be seen by rotating one around. Notice that if the one on the left is rotated 180 degrees, then the Z group will be facing backward, into the paper. To make one like the other, two of the bonds must first be broken and then remade.

If any carbon atom has four different groups attached to it, then it is an **asymmetric carbon**, and the molecule and its mirror image are called **isomers**. Isomers are two or more molecules that have the same molecular formula but different arrangements of atoms in space. This particular kind of isomer, whose members are mirror images of one another, is called an **enantiomer**, or **optical isomer**, and is one form of **stereoisomer**. Many biological molecules are enantiomers, and usually only one of the two forms (and not its mirror image) is biologically active, often because enzymes will not recognize the mirror image.

Geometric isomers, a second kind of stereoisomer, involve double bonds between carbon atoms. There is no free rotation of parts of the molecule around a double bond. The two forms of geometric isomers are called *cis* (when groups are on the same side of the double bond; see Figure 2.1e) and *trans* (when the groups are on the opposite side; see Figure 2.1f). A *cis-trans* isomer called **retinal** plays an essential role in our ability to see.

Structural isomers are a third kind of isomer, and perhaps the easiest to understand. Figures 2.1c and 2.1d show structural isomers, which share the same chemical formula but differ in the arrangement of covalent bonds among their atoms.

a)

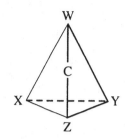

b)

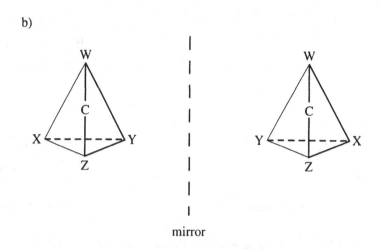

mirror

Figure 2.2 Carbon bonding structure. a. When four atoms (or groups of atoms) are linked to a single carbon atom (C), the shape is that of a tetrahedron. A tetrahedron is a three-dimensional structure with four triangular sides. The four atoms or groups are arbitrarily shown as W, X, Y, and Z. With the carbon atom in the middle of an imaginary tetrahedron, each group will be at one of the corners of the tetrahedron. b. Asymmetric carbon atoms can have enantiomers, which are mirror images of each other. Enantiomers are one form of isomer.

Topic Test 4: Carbon

True/False

1. Structural formulas give more detail about a molecule than do chemical formulas.

2. The fact that carbon atoms form four covalent bonds contributes to the diversity of chemicals so necessary for life.

3. Geometric isomers are a form of stereoisomer.

Multiple Choice

4. Which of the following has a *cis-trans* (geometric) isomer?

$$\text{a. } H_2N-\overset{\overset{\displaystyle H}{|}}{\underset{\underset{\displaystyle CH_3}{|}}{C}}-COOH \qquad \text{b. } \overset{\overset{\displaystyle H_3C \quad CH_3}{|\qquad|}}{\underset{\underset{\displaystyle H \quad H}{|\qquad|}}{C=C}} \qquad \text{c. } H_3C-\overset{\overset{\displaystyle H}{|}}{\underset{\underset{\displaystyle H}{|}}{C}}-\overset{\overset{\displaystyle H}{|}}{\underset{\underset{\displaystyle OH}{|}}{C}}-H$$

$$\text{d. } H_2N-\overset{\overset{\displaystyle H}{|}}{\underset{\underset{\displaystyle CH_3}{|}}{C}}-COOH \qquad \text{e. } HS-\overset{\overset{\displaystyle H}{|}}{C}=O$$

5. A molecule and its mirror image are not the same.
   a. The molecule could be H_2O.
   b. They could be enantiomers or optical isomers.
   c. They are isotopes.
   d. They are geometric isomers.

Short Answer

6. Why do carbon atoms have enantiomers?

Topic Test 4: Answers

1. **True.** The actual bonding arrangement among the atoms is shown in a structural formula, but not in a chemical formula. Some organic molecules have the same chemical formula but different structural formulas.

2. **True.** This property of carbon atoms allows complicated structures to be built from a few atomic building blocks.

3. **True.** Geometric isomers and enantiomers are both stereoisomers.

4. **b.** Geometric isomers require double bonds between one or more pairs of carbon atoms.

5. **b.** Do notice (incorrect) answer c—don't confuse isomers, discussed in this chapter, with isotopes, which we discussed in Chapter 1.

6. Carbon atoms have enantiomers because of their tetrahedral shape. The four atoms or groups attached to a carbon atom are not all in a plane.

APPLICATION: ACID RAIN

Acid rain is thought to be caused by the emissions of pollutants from factories, power plants, autos, and so on. Such substances as nitrates and sulfates are emitted by such industrial activity and contribute to acidity in rain water.

Normal rain is slightly acidic, with a pH of about 5.6. This is because of naturally occurring chemicals, such as carbon dioxide, in the air. As the rain falls, it collects the carbon dioxide, and that can decrease the pH: $CO_2 + H_2O \rightarrow H_2CO_3$, which is a weak acid.

Acid rain can have a pH < 4.2. That acidity can reduce the pH in streams and rivers, ponds, and lakes. What is the ratio of acid in acid rain to that in normal rain? We merely need to convert pH to actual concentrations. Using the formula given earlier in the chapter, $[H^+] = 10^{-pH}$, so the ratio is $10^{-4.2}/10^{-5.6}$, which is equal to 0.0000631/0.0000025, or 25.2 times as much acid! Such a high amount of acid can damage the living organisms that live in the waters receiving acid rain. Laws in the United States intend to reduce the amount of acid rain by increasing energy efficiency, using cleaner fuels, and reducing emissions.

Chapter Test

True/False

1. Tritium is an atom with one proton and two neutrons in its nucleus. It must be an isotope of hydrogen.

2. Tritium has a half-life of about 12.5 years. After 37.5 years, about one-eighth of the initial amount of tritium in a sample will still be present.

Multiple Choice

3. A solution has a pH of 3. The concentration of hydrogen ions in the solution is
   a. 3 moles/liter.
   b. 10^3 moles/liter.
   c. 10^{-3} moles/liter.
   d. 3 grams/liter.
   e. $\frac{1}{3}$ gram/liter.

4. Concerning pH,
   a. the concentration of hydroxyl ions is greater than that of hydrogen ions in a solution at pH 5.
   b. a solution at pH 2 has more hydroxyl ions than a solution at pH 3.
   c. neutral solutions have a pH of 1.0.
   d. a solution with a pH of 2 has more hydrogen ions than a solution with a pH of 3.
   e. all of the above are correct.

5. A substance easily dissolves in water. Which of the following describes characteristics that the substance is likely to have?
   a. Hydrophobic and uncharged
   b. Hydrophilic and polar
   c. Nonpolar
   d. Both a and c
   e. All of the above

6. The atomic mass of an element can be approximated by adding the following numbers:
   a. The number of protons plus electrons
   b. The number of protons plus neutrons
   c. The number protons, plus neutrons, plus electrons
   d. The number of different isotopes of the atom

7. The difference between ^{12}C and ^{14}C is that
   a. ^{14}C has two more electrons.
   b. ^{14}C has two more protons.
   c. ^{14}C has two more neutrons.
   d. both a and c are correct.
   e. all of the above are correct.

Chapter Test Answers

1. **T** 2. **T** 3. **c** 4. **d** 5. **b** 6. **b** 7. **c**

Check Your Performance:

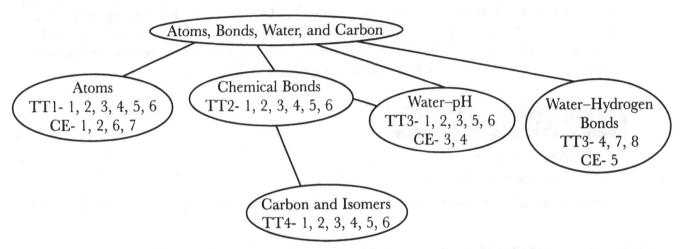

Key: TT = Topic Test; CE = Chapter Exam. Numbers indicate exam questions. Some questions are listed more than once if they refer to more than one topic.

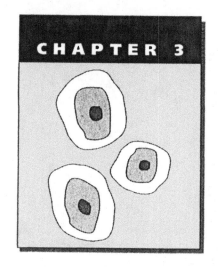

CHAPTER 3

Molecules and Macromolecules of Life

While a large number of different molecules are found in living organisms, knowledge of a few of them is sufficient for understanding much of basic biology. Only a few different kinds of molecules are used to build almost all of the important structures in living organisms. In this chapter we first examine some of the parts of molecules, so-called **functional groups**, that are most often found in the molecules of living organisms. Then we discuss the major kinds of molecules found in living organisms: carbohydrates, lipids, amino acids, and nucleotides. We also examine the common types of **macromolecules** (large molecules) that are formed by linking smaller molecules together.

ESSENTIAL BACKGROUND

- Elements (Chapter 1)
- Chemical bonds (Chapter 2)

TOPIC 1: FUNCTIONAL GROUPS

KEY POINTS

✓ *What are the major, common parts of molecules found in living organisms?*

✓ *What are the properties of the major functional groups?*

We will see the same, few clusters of atoms appearing again and again as we examine the molecules that make up living organisms. These are referred to as **functional groups**, as they play different functions in molecules. Seven such functional groups are described in **Table 3.1**.

Hydroxyl groups are abundantly found in carbohydrates, amino acids, and nucleotides. Linked to a hydrocarbon, a hydroxyl group forms an alcohol. **Carboxyl** groups are hydrogen ion donors. They are weak acids and at pH 7, tend to be in charged form, as shown above. **Amino** groups are found in all amino acids; they are weak bases and tend to attract an additional hydrogen ion at pH 7. **Carbonyl** groups contain a carbon atom doubly bonded to an oxygen. They come in two forms. In **aldehydes**, the carbon is on the end of the molecule, with hydrogen bonded to the same carbon as is linked to the double-bonded oxygen. In **ketones**, the carbon with the doubly bonded oxygen is not on the end, but has organic groups attached on both sides. **Phosphate** groups are abundant in nucleic acids. They are medium-strength acids, and so are charged at normal pH. A **sulfhydryl** group is found in the amino acid cysteine. In proteins, pairs of cysteine residues can link to one another, forming **disulfide bridges** (—S—S—),

Table 3.1 Functional Groups			
NAME	**STRUCTURAL FORMULA**	**USUAL FORM**	**NATURE OF GROUP**
Hydroxyl (—OH)	—O—H		Polar
Carboxyl (—COOH)	$\begin{array}{c}-C{=}O\\ \vert\\ O{-}H\end{array}$	$\begin{array}{c}-C{=}O\\ \vert\\ O^-\end{array}$	Polar and weak acid
Amino (—NH₂)	$\begin{array}{c}-N{-}H\\ \vert\\ H\end{array}$	$\begin{array}{c}H\\ \vert\\ -N{-}H^+\\ \vert\\ H\end{array}$	Polar and weak base
Carbonyl			
Aldehyde	$\begin{array}{c}R{-}C{=}O\\ \vert\\ H\end{array}$		Polar
Ketone	$\begin{array}{c}R{-}C{=}O\\ \vert\\ R'\end{array}$		Polar
Phosphate	$\begin{array}{c}O{-}H\\ \vert\\ -O{-}P{=}O\\ \vert\\ O{-}H\end{array}$	$\begin{array}{c}O^-\\ \vert\\ -O{-}P{=}O\\ \vert\\ O^-\end{array}$	Charged
Sulfhydryl	—S—H		Polar
Methyl	$\begin{array}{c}H\\ \vert\\ -C{-}H\\ \vert\\ H\end{array}$		Nonpolar

which help to hold the protein in a particular configuration. **Methyl** groups are found in abundance in lipid molecules and are the only functional group that is nonpolar. Molecules containing a large number of methyl groups, or linked methyl groups, are **hydrocarbons** and tend to be insoluble in water, or hydrophobic.

Much of the diversity of forms and functions in living organisms results from these functional groups being combined in different ways to form a number of molecules. Some of these molecules, in turn, are used as building blocks in the assembly of macromolecules. In living organisms, more complicated structures are built from simpler components: Linked atoms form functional groups, linked functional groups form molecules, and molecules form macromolecules. This theme will continue in later chapters as we learn about subcellular structures built from macromolecules, cells made up of subcellular structures, tissues and organs made up of cells, organisms composed of tissues and organs, populations and communities of organisms, and ecosystems based on communities and their environments.

Topic Test 1: Functional Groups

True/False

1. Sulfhydryl groups are polar.

2. Phosphate groups are found in nucleic acids, such as DNA.

3. A ketone is a kind of carbonyl group.

Multiple Choice

4. Which of the following molecules is nonpolar?

             H   H
                   |
a. H_3C—C—CH_3 b. H_2O c. H_2N—C—OH
             | |
            CH_3 CH_3

             H   OH
             | |
d. H_2N—C—C=O e. H_2N—C—C=O
             | | |
             H   H OH

5. Which of the following functional groups is a weak base?
   a. Aldehyde
   b. Ketone
   c. Carboxyl
   d. Sulfhydryl
   e. Amino

6. Which of the following functional groups is nonpolar?
   a. Carboxyl
   b. Carbonyl
   c. Amino
   d. Methyl
   e. Phosphate

Short Answer

7. Why are methyl groups nonpolar?

Topic Test 1: Answers

1. **True.** Sulfur is electronegative, and so produces a slight positive charge on the hydrogen, leading to its being polar.

2. **True.** We will see that phosphate groups, alternating with sugar groups, form the backbones of DNA molecules.

3. **True.** Both aldehydes and ketones are carbonyl groups.

4. **a.** Notice that answer a consists mostly of methyl groups, and is nonpolar. All of the other molecules have polar groups present.

5. **e.** Only the amino group tends to gain a hydrogen ion, and thus is a weak base.

6. **d.** All of the other groups are polar or charged.

7. Methyl groups are nonpolar because, as pointed out in the first chapter, bonded carbon and hydrogen atoms share pairs of electrons equally. As a result, there is an equal charge distribution between carbon and hydrogen atoms.

TOPIC 2: CARBOHYDRATES

KEY POINTS

✓ *What is the general chemical formula for carbohydrates?*

✓ *Name two important pentose sugars.*

✓ *What is the difference between glycogen and cellulose?*

✓ *What are condensation (dehydration) synthesis and hydrolysis?*

Carbohydrates have roughly equal numbers of carbon and oxygen atoms, and twice that number of hydrogen atoms. Thus, an approximate chemical formula for a typical carbohydrate is $(CH_2O)_n$, where "n" is an integer (3, 4, 5, etc.). Notice that this formula can be viewed as a carbon ("carbo") plus water ("hydrate"). A typical simple carbohydrate will consist of a string of linked carbon atoms, with hydrogen and hydroxyl groups linked to all but one of the carbon atoms, which will contain a carbonyl group. Most carbohydrates contain asymmetric carbons, and thus have isomers, usually more than one.

One very important carbohydrate is **glucose** $(C_6H_{12}O_6)$, which consists of a chain of six linked carbon atoms **(Figure 3.1)**. This **monosaccharide**, or simple sugar, when placed in water actually forms a ring, as shown in Figure 3.1. Two different rings can be formed because an asymmetric carbon is created when the ring forms. The two ring versions of glucose are called **alpha** and **beta**.

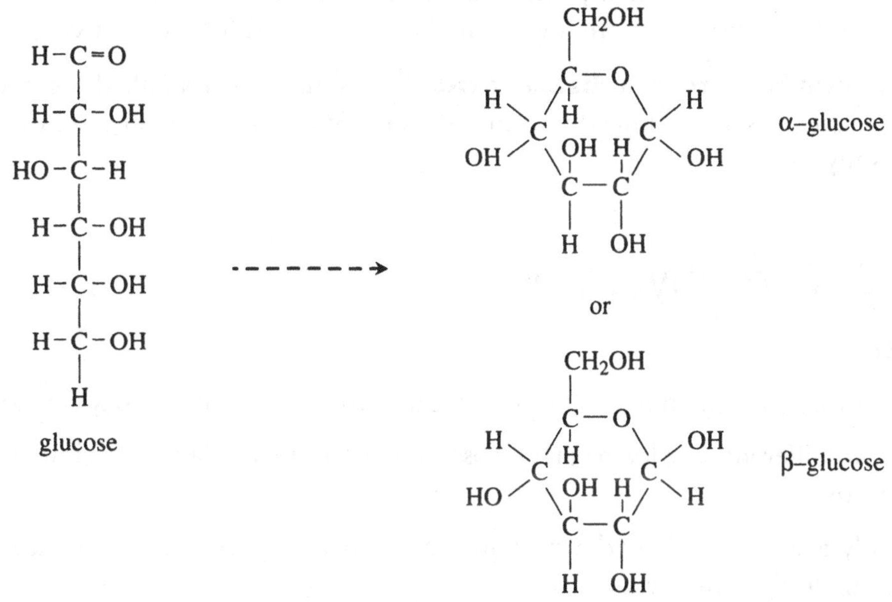

Figure 3.1 Glucose. Glucose molecules are carbohydrates containing six carbon atoms. In water, a glucose molecule can spontaneously form one of two ringed compounds, alpha or beta, which are enantiomers. When the ring forms, carbon atom 1 (part of an aldehyde group) links to the oxygen atom attached to carbon atom 5 in the chain. As the ring forms, carbon atom 1 becomes an asymmetric carbon, and alpha and beta are the two possible isomers.

Many copies of a simple sugar can be linked to form carbohydrate macromolecules, which are called **polysaccharides**. For example, many molecules of glucose in the alpha form are linked together to make **glycogen**, a macromolecule used for energy storage in living organisms. **Starch** is another polysaccharide made by linking together glucose molecules in the alpha ring form. In contrast, **cellulose**, a major polysaccharide of plant cell walls, is made by linking together glucose molecules in the beta ring form.

In all cases, the linkage of the glucose molecules involves the removal of a water molecule (a hydroxyl group from one glucose and a hydrogen from the hydroxyl of the next glucose are removed to form one water molecule as the next monomer gets linked to a growing chain). This removal of water as a chemical linkage occurs is called **condensation (or dehydration) synthesis**. We will see that condensation synthesis is also used to link other kinds of monomer units into other kinds of macromolecules. This macromolecule-building reaction usually requires energy.

The opposite process involves the splitting of the glucose units out of the chain, and this is called a **hydrolysis** reaction because water is split and added to the divided molecules—a hydroxyl group to one and a hydrogen to the other.

Not all carbohydrate macromolecules are easily hydrolyzed. The seemingly little difference between alpha and beta forms of glucose makes a big difference in the ability of most organisms to break down the macromolecules that are formed. Many animals are able to hydrolyze glycogen, starch, and other polysaccharides made from the alpha ring form of glucose. However, only a few microorganisms are able to digest cellulose, the macromolecule made from the beta ring form. Cows make use of such microorganisms during digestion to hydrolyze cellulose in grasses.

Two other important monosaccharides are **ribose** and **deoxyribose**. Both of these ribose sugars consist of five-carbon chains, which also form rings in solution. Ribose and deoxyribose are parts of the nucleotides that are used to make RNA and DNA, respectively.

Finally, table sugar is **sucrose**, a **disaccharide** (two simple sugars linked together), which consists of a molecule of glucose linked to a molecule of **fructose**, another six-carbon carbohydrate commonly found in fruits.

Topic Test 2: Carbohydrates

True/False

1. Carbohydrates contain about twice as many hydrogen atoms as they do oxygen atoms.

2. The two different ring forms of glucose, alpha and beta, play quite similar roles in living organisms.

3. Hydrolysis is the breaking down of molecules involving the addition of a split water molecule to the two parts.

Multiple Choice

4. Which of the following functional groups is found in glucose?
   a. Hydroxyl
   b. Carboxyl

c. Amino
  d. Methyl
  e. Phosphate

5. In what way(s) are glycogen and cellulose the same?
  a. They both are made from the alpha form of glucose.
  b. They both are carbohydrate polymers (polysaccharides).
  c. They both are made by condensation synthesis.
  d. Both b and c are correct
  e. All of the above are correct.

Short Answer

6. Do you expect glucose to be very soluble in water? Why?

Topic Test 2: Answers

1. **True.** Carbohydrates have a chemical formula close to $(CH_2O)_n$.

2. **False.** Their very different roles are described above.

3. **True.** Hydrolysis is the opposite of condensation synthesis.

4. **a.** There are five hydroxyl groups in glucose, and none of the others that are listed.

5. **d.** They are not both made from alpha glucose.

6. Yes, glucose should be very soluble in water since it has so many polar functional groups present.

TOPIC 3: LIPIDS

KEY POINTS

✓ *Lipids contain a lot of what two elements?*

✓ *Why are lipids expected to be hydrophobic?*

✓ *What are fatty acids, fats, and phospholipids?*

✓ *What is the shape of cholesterol, and what is it used for?*

Lipids consist predominantly of carbon and hydrogen. Lipids can contain a few oxygen, nitrogen, and phosphorus atoms, but they are in low abundance relative to carbon and hydrogen. Because the abundant carbon and hydrogen atoms are mostly linked to one another, almost all of the bonds in a lipid are nonpolar, and thus lipids tend to be hydrophobic (Chapter 2), and do not dissolve appreciably in water.

Perhaps the simplest lipids are **fatty acids**, which are long hydrocarbon chains—perhaps 15 to 20 linked carbon atoms with only hydrogen atoms attached, ending in a (usually ionic) carboxyl group. **Saturated** fatty acids have only single bonds between all carbon atoms in the fatty acid molecule. **Unsaturated** fatty acids have one double or triple bond, and **polyunsaturated** fatty acids more than one double or triple bond, and thus are "unsaturated" in terms of the number of hydrogen atoms that are a part of the fatty acid (a double bond has two fewer hydrogen atoms, and a triple bond has four fewer hydrogen atoms). **Polyunsaturated** fatty acids tend to have lower melting points and tend to be liquid (oil) rather than solid at room temperature. Double bonds between carbon atoms produce "kinks" in the fatty acid chain.

Fat molecules contain three fatty acid chains, all linked to a three-carbon glycerol molecule. The fatty acid chains are added to the glycerol by condensation reactions, one fatty acid to each carbon in the glycerol. Fats are used for insulation, padding, and as a storage form for chemical energy in animals.

Phospholipids also contain glycerol, but only two fatty acid chains are attached. The third carbon on the glycerol is occupied by a phosphate group. Additional functional groups can be added to the phosphate group, forming different kinds of phospholipids. The phospholipid can be viewed as consisting of a hydrophilic "head" group, containing the phosphate group, plus two, hydrophobic, fatty acid "tails." We will see that phospholipids play a very important role in the structure of cell membranes.

Not all lipids contain fatty acid chains. **Cholesterol** is a lipid that consists of four, interlinked hydrocarbon rings, with a few functional groups attached to the rings. While the negative role of cholesterol in heart disease often is emphasized, cholesterol plays important roles both as a membrane constituent and as a precursor for steroid hormones, sex hormones, vitamin D, and bile acids.

Topic Test 3: Lipids

True/False

1. Fatty acids are made from glycerol.

2. Fats are made, in part, from glycerol.

3. Phospholipids are important in membranes.

4. Fats contain two fatty acid chains.

Multiple Choice

5. Lipids
   a. are mostly hydrophilic.
   b. are very soluble in water.
   c. contain many polar groups.
   d. are hydrophobic, and not water soluble.
   e. Answers a, b, and c are correct.

6. A saturated fatty acid
   a. contains one carbon-carbon double bond.
   b. contains two carbon-carbon double bonds.
   c. contains many carbon-carbon double bonds.
   d. does not contain double bonds.

Topic Test 3: Answers

1. **False.** Glycerol is combined with fatty acids to make fats.

2. **True.** Fats are made by linking three fatty acids to glycerol.

3. **True.** We will see that phospholipids make up about half of a typical membrane.

4. **False.** Fats contain three such chains.

5. **d.** Lipids are (mostly) nonpolar molecules.

6. **d.** Unsaturated and polyunsaturated fatty acids contain double bonds.

TOPIC 4: AMINO ACIDS AND PROTEINS

KEY POINTS

✓ *What are amino acids?*

✓ *How many different kinds of amino acids do living organisms use to make proteins?*

✓ *What are primary, secondary, tertiary, and quaternary structures of proteins, and what bonds or forces hold these structures in place?*

Amino acids are the building blocks for proteins. The general structure for an amino acid is

$$H_2N-\underset{\underset{R}{|}}{\overset{\overset{H}{|}}{C}}-COOH$$

where "R" is an organic group. At pH 7, the structure becomes

$$^+H_3N-\underset{\underset{R}{|}}{\overset{\overset{H}{|}}{C}}-COO^-$$

Notice that at this pH, each amino acid has both a positive and negative charge, as well as whatever charge might come from the R group.

Twenty different amino acids are used to make proteins. The same 20 are used in all organisms. These 20 amino acids can be modified after the proteins are formed, so a large number of different kinds of amino acids are found in the proteins of living organisms.

By examining the general structure of amino acids, we can see where the name "amino acid" came from—there is a central carbon that has an amino group and an acid (carboxyl) group linked to it. The central carbon also has a hydrogen and a generalized R group attached. In this case, the R group can be either an organic group or another hydrogen. In the latter case, the amino acid is glycine. The 19 amino acids other than glycine have a central carbon atom that is asymmetric (Chapter 2). Thus, there are two isomers for each of the 19 amino acids. Only one of the two (the "L" form) is used to make proteins.

The R groups on amino acids are called **side groups**, and they determine the nature of the amino acid and its role in proteins. Some side groups are charged, some are uncharged but polar, and some are nonpolar.

Polypeptides are linear chains of amino acids. In polypeptides, the amino acids are linked together like beads on a string. Each protein consists of one or a few polypeptides. The amino acids in the chain are covalently bonded to each other by links between the amino group on one amino acid and the acid (carboxyl) group on the next:

$$\overset{\displaystyle H}{\underset{\displaystyle R_1}{^+H_3N-C-COO^-}} + \overset{\displaystyle H}{\underset{\displaystyle R_2}{^+H_3N-C-COO^-}} \rightarrow \overset{\displaystyle H\quad O\quad H}{\underset{\displaystyle R_1\quad H\ R_2}{^+H_3N-C-C-N-C-COO^-}} + H_2O$$

Notice that a water molecule has been removed, making this a condensation (dehydration) synthesis. Also notice that the carbon atom in the carboxyl group from the first amino acid is linked directly to the nitrogen from the amino group of the second amino acid. This is called a **peptide bond**. All of the amino acids in a protein chain are linked by such peptide bonds. You should be able to see how a chain can be built by adding a third amino acid to the pair shown above, and then a fourth, and a fifth, and so on. Finally, notice that the charges on the "backbone" of the polypeptide are eliminated as the peptide bonds are formed. Just one charge at each end of the polypeptide remains, plus whatever charges are present on the side groups. Therefore, the side groups play a big role in determining whether a local region of the protein is polar or nonpolar, charged or uncharged, and so on.

Since any one of 20 different kinds of amino acids can occupy each position on the polypeptide chain, and since such chains can be hundreds of amino acids long, many different kinds of proteins are possible. The typical size of a protein is about 50,000 daltons. With a typical amino acid having a size of about 120 daltons, we calculate that there are about 400 amino (50,000 divided by 120) acids in a typical protein.

The **primary structure** of a protein, or polypeptide chain, is the amino acid sequence. The **secondary structure** is defined as the local folding of the chain. Hydrogen bonds play a crucial role in local folding. Examples of secondary structure include the **alpha-helix** and the **beta-pleated sheet**. **Tertiary structure** is the more global folding of the polypeptide chain. Four factors or forces hold the tertiary structure in place: 1) hydrogen bonds; 2) nonpolar, hydrophobic groups that often come together; 3) ionic interactions among charged side groups; and 4) bonding of pairs of cysteine residues, which contain sulfhydryl groups. The bonding of the pairs of cysteine residues to each other results in the formation of disulfide (—S—S—) bridges between different parts of the chain. If there is more than one polypeptide chain in a protein, then **quaternary structure** holds the chain together. The forces involved in quaternary structure are the same as for tertiary structure. For the roles of proteins in cells, see the web page.

Topic Test 4: Amino Acids and Proteins

True/False

1. Twenty different kinds of amino acids are used by cells to make proteins.

2. Amino acids contain carboxyl groups and amino groups.

Multiple Choice

3. Which of the following is (are) involved in generating tertiary structure of proteins?
   a. Hydrogen bonds
   b. Ionic bonds
   c. Hydrophobic interactions
   d. Disulfide bonds
   e. All of the above

4. Which of the following is most likely to be an amino acid?
   a. $C_7H_{14}O_7$
   b. $C_3H_{11}O_2N$
   c. $C_4H_{13}N_2$
   d. $C_4H_{11}N_2P$

5. Which of the following is (are) an example of secondary structure in proteins?
   a. Hydrogen bonds
   b. Glycine
   c. Alpha-helix
   d. Peptide bond

Short Answer

6. A typical protein is 400 amino acids long. How many different kinds of proteins of this length are possible?

Topic Test 4: Answers

1. **True.** The same 20 appear to be used in all living organisms.

2. **True.** That is why they are called amino acids.

3. **e.** All four of these forces or interactions influence tertiary structure.

4. **b.** Different amino acids can have different chemical formulas, depending on the nature of the side group. However, the presence of an amino group and a carboxyl group requires the presence of C, H, N, and O in all amino acids. The other answers did not have all four of these elements.

5. **c.** The alpha-helix is one kind of secondary structure. It was proposed as such by Linus Pauling, who was awarded two Nobel prizes during his life.

6. In each position there are 20 possibilities, one for each of the 20 amino acids. The following reasoning can be used: For two linked amino acids there are $20 \times 20 = 20^2 = 400$ possibilities; for three linked amino acids there are $20 \times 20 \times 20 = 20^3 = 8,000$ different triplets; . . . so for 400 linked amino acids there would be 20^{400} possibilities. Try punching that into a typical hand calculator and you will get a big "E" (error) on the scoreboard because the number is so large. In fact, if you were to try to make just one copy of each, you would run out of atoms on earth before you were finished (actually you'd run out of time first!). Nature has tried a number of these possible proteins, and others of different lengths than 400 amino acids, and has kept the ones that are useful by the process of natural selection, which we will discuss in Chapter 16.

TOPIC 5: NUCLEOTIDES AND NUCLEIC ACIDS

KEY POINTS

✓ *What three components are linked to make a nucleotide?*

✓ *What is ATP and what role does it play in cells?*

✓ *What are the building blocks from which the nucleic acids, DNA and RNA, are composed?*

✓ *What are the chemical differences between DNA and RNA?*

Five **nitrogenous bases** are commonly found in nucleic acids. Three of these are **pyrimidines—cytosine (C), thymine (T),** and **uracil (U).** Two are **purines—adenine (A)** and **guanine (G).** All of these nitrogenous bases consist of rings of linked carbon and nitrogen atoms. Pyrimidines contain one such ring; purines contain two.

Nucleosides contain a nitrogenous base linked to a pentose (five-carbon) sugar. Two kinds of pentose sugars are normally used: ribose in RNA and deoxyribose in DNA. An example of a nucleoside is adenosine, which consists of an adenine base linked to a ribose sugar.

Nucleotides consist of nucleosides whose sugar is linked to a phosphate group as well as a nitrogenous base. An example of a nucleotide is adenosine monophosphate, or AMP, which consists of the following linked groups: adenine-ribose-phosphate.

Nucleic acids are made of linked nucleotides. There are two kinds of nucleic acids—**DNA** and **RNA**—and they have some features in common. Both consist of linear chains of linked nucleotides. A short chain might be about 70 nucleotides long; longer chains can contain millions of nucleotides. In the nucleic acids, the phosphate on one nucleotide is linked to the sugar on the next nucleotide, forming a repeating, sugar-phosphate backbone, with the nitrogenous bases hanging out like the side groups of amino acids do in proteins. The differences between DNA and RNA are as follows:

Nucleic Acid	Bases	Pentose Sugar	Usual Number of Strands
DNA	A,T,G,C	Ribose	Two
RNA	A,U,G,C	Deoxyribose	One

There are other uses for nucleotides in cells. In a specially modified form they are used as an energy source that allows living cells to do work, from driving a chemical reaction to moving packages around in cells. This modified form is called a **nucleoside triphosphate**. Such molecules have three linked phosphate groups. The most commonly used one in cells is **adenosine triphosphate**, or **ATP**, whose structure is adenine-ribose-phosphate-phosphate-phosphate. ATP is a source of energy (so-called free energy, as we will call it in Chapter 6) that living cells use to do work. The molecule releases energy that can be used to do work when it is hydrolyzed, splitting off one or two phosphates from the triphosphate.

$$ATP + H_2O \rightarrow ADP + P_i \quad \Delta G = -7.3 \, kcal/mole$$

where ADP is adenosine diphosphate; P_i is inorganic phosphate—one of our functional groups; and ΔG is a measure of the amount of energy released by the hydrolysis of the ATP. When ΔG is negative, it indicates that free energy is released by the reaction. Triphosphates also are used to make DNA and RNA; only then, two phosphate groups are split off, leaving a nucleotide that links to the chain. For a couple of "secrets of life," see the web page.

Topic Test 5: Nucleotides and Nucleic Acids

True/False

1. Nucleosides contain base-sugar-phosphate.

2. ATP contains adenine-sugar-phosphate-phosphate-phosphate.

3. DNA contains the four bases: adenine, thymine, uracil, and guanine.

Multiple Choice

4. Which of the following shows the structure of a nucleotide?
   a. Phosphate-pentose sugar-base
   b. Adenine-base-phosphate
   c. Phosphate-adenine-base
   d. Phosphate-lipid-pentose sugar
   e. Sugar-lipid-base

5. Which of the following is found in RNA, but not in DNA?
   a. Ribose
   b. Uracil
   c. Deoxyribose
   d. Thymine
   e. Both a and b

6. ATP
   a. contains three phosphate groups.
   b. is a source of energy for doing work in cells.
   c. contains adenine.
   d. is a nucleoside triphosphate.
   e. All of the above are correct.

Short Answer

7. How can a sequence of nucleotides in a nucleic acid contain information?

Topic Test 5: Answers

1. **False.** Nucleotides contain base-sugar-phosphate.

2. **True.** The sugar is ribose.

3. **False**. DNA contains adenine, thymine, cytosine, and guanine.

4. **a.** Phosphate is linked to a ribose or deoxyribose, which in turn is linked to a nitrogenous base.

5. **e.** RNA contains ribose (DNA contains deoxyribose) and uracil (DNA contains thymine).

6. **e.** All are correct about ATP.

7. Since there are four different bases present along the strand of a DNA molecule, it is like a four-letter alphabet, A, G, C, and T. In genetic "language," as we will see in Chapter 10,

there actually are "words," each three letters long. Just as human language contains information, so can this genetic language. In the genetic language, most of the words specify particular amino acids.

APPLICATION: THE CALORIC CONTENT OF FOODS

We have gained enough knowledge about the structure of different kinds of molecules to understand why certain foods contain more calories than others. Calories are a measure of the heat energy present in the bonds of food molecules. Such heat energy can be released by burning the organic molecules in the presence of oxygen, producing water and carbon dioxide. Of course, living organisms do not actually burn food (except some of us do so by mistake on the stove!), but living organisms do extract the energy and do produce water and carbon dioxide, as we will see in Chapter 7. As carbon dioxide and water are produced, the carbon-carbon and carbon-hydrogen bonds in the molecules in food are converted to carbon-oxygen and hydrogen-oxygen bonds.

It is commonly known that fatty foods contain more calories per weight than do foods that gain their calories primarily from carbohydrates or proteins. By observing the difference in chemical formulas between lipids as opposed to carbohydrates and proteins, we can gain insight into why. Lipids are predominantly hydrocarbons, and so have many carbon-carbon and carbon-hydrogen bonds. In fact, they have about twice as many such bonds as do carbohydrates and proteins. That explains why lipids also contain about twice the energy, by weight, of carbohydrates and proteins. In animals, fat can be a useful, high-density storage form for energy.

Chapter Test

True/False

1. A peptide bond links two carbohydrates together.

Multiple Choice

2. Lipids differ from carbohydrates in that lipids
   a. are found in cells only when food is not abundant.
   b. are rarely found in cells.
   c. contain less oxygen per unit of carbon.
   d. are linked sugar molecules.
   e. do not contain carbon or hydrogen.

3. Individual polypeptide chains locally bend or fold into coils or sheets. This is the
   a. zero-order structure.
   b. primary structure.
   c. secondary structure.
   d. tertiary structure.
   e. quaternary structure.

4. The following molecule is
   a. a steroid.
   b. an amino acid.
   c. cholesterol.
   d. a nitrogenous base.
   e. a fatty acid.

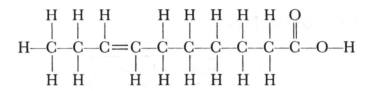

5. The synthesis of carbohydrate macromolecules and of proteins share which of the following features?
   a. Dehydration (condensation) synthesis
   b. The presence of sugar molecules
   c. Formation of peptide bonds
   d. Amino group linkages
   e. All of the above

6. Of the following side (R) groups from amino acids, the one which has the *least* affinity for water is
   a. $-CH_2-CH_2-CH_2-NH-C=NH_2$ b. $-CH_2CH_2-CH=O$
   (with NH_2 below)   (with NH_2 below)
   c. $-CH_2-OH$ d. $-CH-CH_2-CH_3$
                      (with CH_3 below)

7. A chemical with the formula $C_{11}H_{22}O_2$ probably is a(n)
   a. nucleic acid.
   b. protein.
   c. carbohydrate.
   d. fatty acid.
   e. amino acid.

8. Can a carbohydrate be covalently bonded to a nitrogenous base?
   a. Yes
   b. No
   c. Only if a phosphate group is present

9. Fat molecules
   a. contain many linked sugars.
   b. are mostly hydrocarbon.
   c. are water soluble.
   d. All of the above are correct.
   e. None of the above are correct.

Short Answer

10. What are the four forces, or factors, responsible for tertiary structure in proteins?

Chapter Test Answers

1. **F** 2. **c** 3. **c** 4. **e** 5. **a** 6. **d** 7. **d** 8. **a** 9. **b**

10. The four forces are hydrogen bonds, hydrophobic interactions, ionic bonds, and disulfide bridges.

Check Your Performance:

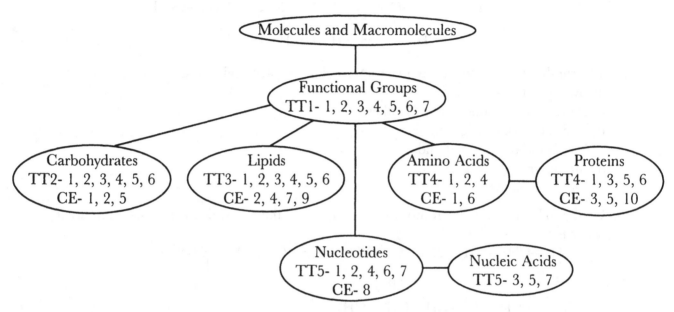

Key: TT = Topic Test; CE = Chapter Exam. Numbers indicate exam questions. Some questions are listed more than once if they refer to more than one topic.

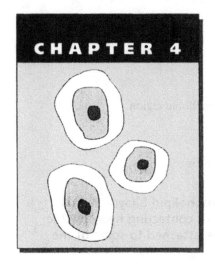

CHAPTER 4

Membrane Structure and Function

Membranes are the barriers between living cells and their environments. Membranes also subdivide eukaryotic cells into compartments. Membranes allow the passage of particular materials into, and out of, cells. We begin by considering the general structure of membranes, and then consider how materials pass through membranes.

ESSENTIAL BACKGROUND

- Phospholipids (Chapter 3)
- Proteins (Chapter 3)

TOPIC 1: MEMBRANE STRUCTURE

KEY POINTS

✓ *What are the major chemical constituents of membranes?*

✓ *Where are the hydrophobic and hydrophilic regions in membranes?*

✓ *What is a lipid bilayer?*

✓ *What is the fluid mosaic model of the membrane?*

Membranes in living organisms are about half lipid, half protein, and less than 10% carbohydrate. The most abundant lipids are phospholipids. Membranes contain two phospholipid layers (a **phospholipid bilayer**), with the fatty acid tails of the phospholipids in each layer pointing toward the tails in the other layer. In this way the hydrophobic tails are in contact, and are kept separate from the water solutions on either side of the membrane. **Figure 4.1** shows a cross-section through a membrane. Normally the membrane folds around on itself, forming a surface that fully encloses a volume of liquid, separating it from the liquid on the outer side of the membrane.

Some proteins also are integral parts of membranes. Such **integral membrane proteins** usually are fully embedded in the membrane, or even pass through from one side to the other. Integral membrane proteins have hydrophobic amino acid side groups imbedded in the membrane, in contact with the hydrophobic tails of the phospholipids. The hydrophilic head groups of the phospholipids and the hydrophilic side groups of the proteins have contact with the liquid solutions outside the membrane. This organization minimizes the free energy of the membrane and surrounding water solutions—hydrophobic parts are in contact with other hydrophobic

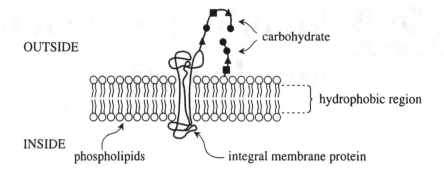

OUTSIDE

carbohydrate

hydrophobic region

INSIDE

phospholipids

integral membrane protein

Figure 4.1 A biological membrane in cross-section, consisting of a phospholipid bilayer, with a hydrophobic interior containing the fatty acid tails of the phospholipids contacting hydrophobic parts of the imbedded integral membrane proteins. Carbohydrates are attached to some of the proteins and phospholipids on one side of the membrane.

parts, hydrophilic with hydrophilic. Carbohydrate groups present in cell membranes usually are bonded to protein or lipid molecules and are on the exposed, outer surface of the cell.

Some other proteins may be more loosely bound to membranes, and these are called **peripheral membrane proteins**. Some of these can also be linked to cytoskeletal components inside the cell, thereby giving stability to the structure of the cell and contributing to possible movement and shape change in cells as well, as we will see in Chapter 5.

The lipids within a layer can move rapidly by one another, and unless held in place, proteins also can move within the plane of the membrane. This is the **fluid mosaic model** of the membrane—a mosaic, or mixing, of different proteins that behave as if drifting in a fluid sea of lipid. The phospholipids cannot flip-flop from one layer to the other without help of some sort. Phospholipids that are polyunsaturated, with "kinks" in their tails due to double or triple bonds, produce more fluid membranes. Saturated fatty acid tails on the phospholipids produce more viscous, less fluid membranes. Cholesterol, another lipid component of membranes, increases membrane thickness and rigidity.

Proteins play multiple roles in membranes.

Topic Test 1: Membrane Structure

True/False

1. Membranes contain lipids, carbohydrates, and proteins.

2. Phospholipids easily flip from one layer in the membrane to the other.

3. Carbohydrate groups are bound to the exterior side of the outer membrane of cells.

Multiple Choice

4. The inner, hydrophobic region of a membrane contains
   a. the carbohydrate tails of phospholipid molecules.
   b. hydrophobic regions of membrane proteins.

c. significant amounts of water.

d. many polar groups.

e. both a and b.

5. Concerning membranes,

a. polyunsaturated phospholipids produce more fluid membranes.

b. cholesterol contributes to membrane rigidity.

c. lipids in a given layer can move by one another.

d. carbohydrates can be found linked to proteins in cell membranes.

e. all of the above are correct.

Short Answer

6. Why is energy a minimum when hydrophobic links with hydrophobic rather than mixing with hydrophilic parts of the membrane?

Topic Test 1: Answers

1. **True.** These are the major constituents of membranes.

2. **False.** They can move within a bilayer, but not easily from one to the other.

3. **True.** They are linked to proteins and lipids.

4. **e.** Notice that hydrophilic, polar groups are not abundantly found in contact with the inner region of membranes.

5. **e.** All of these statements are correct.

6. When hydrophobic entities mix with hydrophilic ones, then the hydrogen bonds, or ionic bonds, among the hydrophilic molecules must break, and this requires additional energy.

TOPIC 2: PASSIVE TRANSPORT ACROSS MEMBRANES

KEY POINTS

✓ *What is diffusion, and why do molecules move down their concentration gradients?*

✓ *What is osmosis? What are hyperosmotic and hypoosmotic solutions?*

✓ *What is an electrochemical equilibrium potential?*

✓ *What is facilitated diffusion?*

Passive transport is the net movement of materials across membranes that does not require any special source of energy. The simplest form of passive transport is **diffusion**, which is the net movement of molecules down their **concentration**, or **chemical**, **gradient**, from high to low concentration. Diffusion is the consequence of random, thermal motion, which will tend to make the concentrations more equal. Thus, if we have a 2 M (M = molar = moles/liter) solution of glucose on one side of a membrane, and a 1 M solution on the other, then with time, more glucose molecules will pass in the direction from the 2 M side to the 1 M side than in the other direction. With time, this diffusion will tend to make the glucose concentrations equal on both sides of the membrane. Of course, we are presuming that glucose is permeable, that is, able to cross the membrane. Since glucose is a polar molecule, it requires special channels or mechanisms for its passage across the membrane. Other, typically nonpolar, substances can diffuse directly across cell membranes. Most membranes also are permeable to water. The mechanism for water's permeability still is being studied.

If a membrane is not permeable to glucose, but is permeable to water, then water will move down *its* concentration gradient—with a net flow, or flux, from the 1 M side (where water is in higher concentration) to the 2 M side. This flux of water is called **osmosis**. The net movement of water can be offset by applying a pressure on the 2 M side, resisting the flux of water. Thus, the concentration difference in water produces an **osmotic pressure**, whose magnitude depends on the difference in concentration of the glucose (and hence water) on the two sides of the membrane.

When comparing the concentrations of particles on two sides of a membrane, such as the cell membrane separating the inside of a living cell from the outside, we have special terms to define three possible situations. If the solution outside of the cell has a concentration of particles (molecules, ions, etc.) that is greater than that on the inside, we say the outside solution is **hyperosmotic**, or **hypertonic**. A hyperosmotic solution will gain water from the cell, as osmosis leads to a loss of water from the cell. In contrast, if the outside solution is **hypoosmotic** or **hypotonic**, the cell will gain water by the diffusion of water from outside to inside. If concentrations are equal on the inside and outside, we speak of the solutions as being **isosmotic**, or **isotonic**. Hyperosmotic solutions cause animal cells to shrink and shrivel. Hypotonic solutions cause animal cells to swell, and can even cause them to burst or lyse.

The presence of charged particles across membranes presents an extra level of complexity because their movements are governed by two different influences—that of diffusion due to concentration gradients and that of attraction or repulsion by the potential (voltage) difference between the two sides of the membrane. Consider a membrane separating solutions of potassium chloride on one side and just water on the other:

$$
\begin{array}{lcl}
 & \text{Membrane} & \\
\text{Side 1} & \text{x} & \text{Side 2} \\
 & \text{x} & \\
H_2O & \text{x} & H_2O \\
 & \text{x} & \\
K^+ & \text{x} & \\
Cl^- & \text{x} & \\
 & \text{x} & \\
 & \text{Membrane} & \\
\end{array}
$$

If both potassium (K^+) and chloride (Cl^-) ions are permeable (can cross the membrane), then both will diffuse down their concentration gradients, and continue to do so, until, at equilibrium, the concentrations of each are equal on the two sides of the membrane.

If only potassium is permeable, it will begin to diffuse from side 1 to side 2, but a charge difference will quickly build, positive on side 2 and negative on side 1, because of the positive charge on each potassium ion. This will produce a **potential (voltage, electrical) gradient** across the membrane. Now we have an electrical gradient as well as a chemical one, which together are called an **electrochemical gradient**. In this case, the electrical gradient, as it grows, opposes the chemical one—the potential across the membrane will begin to reduce the flux of K^+ toward a positively charged side 2. At equilibrium, the following equation holds:

$$E_K = (RT/zF)\ln([K_2]/[K_1])$$

where E is the potential (voltage) across the membrane; R and F are constants; ln is the natural logarithm; T is absolute temperature in degrees Kelvin; z is the charge on potassium, +1; and [ ] indicates concentration. This equation simplifies to: $E_K = (60\,mV)\log_{10}([K_2]/[K_1])$. Thus, a tenfold concentration difference for potassium will result in a potential difference of 60 millivolts at equilibrium. This is called the **equilibrium potential** for potassium, and its magnitude and sign depend on the ratio of concentrations of potassium on the two sides of the membrane. We will make use of equilibrium potentials when we consider how neurons work in Chapter 21.

A special form of diffusion in membranes is called **facilitated diffusion**, which is carrier mediated; that is, membrane proteins help other molecules to cross the membrane. The movement is both ways, and is passive—molecules that exhibit facilitated diffusion run down their concentration gradients only. Facilitated diffusion is a way of getting particular hydrophilic molecules across membranes.

Topic Test 2: Passive Transport

True/False

1. Osmosis is produced by diffusion of water.

2. Diffusion causes a net flux of material down concentration gradients.

3. Facilitated diffusion is able to pump molecules up their concentration gradients.

Multiple Choice

4. Two solutions are separated by a membrane. One contains 1 M glucose; the other contains 2 M glucose.
   a. There will be a net flux of water from the 1 M side to the 2 M side.
   b. There will be a next flux of glucose from the 1 M side to the 2 M side.
   c. If glucose is permeable, then at equilibrium, the 1 M side will contain a higher concentration of water.
   d. If glucose is not permeable, then at equilibrium, there will be equal concentrations of glucose on the two sides of the membrane.
   e. All of the above are correct.

5. An animal cell is placed in a hypoosmotic (hypotonic) solution. The result will be
   a. a shrinking of the cell.
   b. a swelling of the cell, perhaps even a bursting.
   c. no change in the size of the cell.
   d. a swelling of the cell, followed by a shrinking.
   e. no change at first, then a shrinking.

6. A membrane is permeable only to water and H^+ ions. It separates two solutions. Side 1 consists of water and side 2 has some HCl (H^+ plus Cl^-) present. What will happen to the pH as equilibrium is approached? The pH on

a. side 1 will decrease.
b. side 2 will decrease.
c. side 1 will not change.
d. both sides will increase.
e. both sides will decrease.

Topic Test 2: Answers

1. **True.** There is a net flux of water down its concentration gradient—from the side where water concentration is greater to the other side. Water concentration is greater when particle concentration is less.

2. **True.** There are more particles moving from the higher concentration toward the lower than in the other direction.

3. **False.** Facilitated diffusion can only produce a net movement of molecules down their concentration gradient.

4. **a.** Water will move down its concentration gradient.

5. **b.** Water will move into the cell.

6. **a.** Hydrogen ions will flux from side 2 to side 1. As hydrogen ion concentration increases on side 1, pH decreases.

TOPIC 3: ACTIVE TRANSPORT ACROSS MEMBRANES

KEY POINTS

✓ *What is active transport?*

✓ *What is the role of Na-K ATPase?*

✓ *What is co-transport?*

✓ *What are endocytosis and exocytosis?*

Active transport is the movement of materials across membranes that requires a source of energy. Active transport is able to pump materials up their concentration gradients, increasing the difference in concentration across a membrane. The energy can come from ATP hydrolysis (Chapter 3), or a preexisting concentration gradient for an ion can serve as the energy source for active transport of another ion or molecule.

Normal cells have much more potassium inside than outside, and much more sodium outside than inside. Perhaps the best-studied mechanism of active transport is the **sodium-potassium pump (sodium pump, Na-K pump, or Na-K ATPase)**, which is a large, membrane protein responsible for pumping both Na^+ and K^+ up their concentration gradients. The sodium-potassium pump uses energy from the hydrolysis of ATP to move sodium from inside the cell to outside and to move potassium in the opposite direction. Thus, the pump is responsible for maintaining the chemical gradients for both sodium and potassium ions. The pump works in the following cycle:

• Three Na ions from inside the cell bind to the pump protein.

• ATP is hydrolyzed, with the phosphate group binding to the pump protein.

- Pump protein changes conformation.

- Three Na$^+$ ions are released outside of cell.

- Two K$^+$ ions from outside the cell bind to the pump protein.

- Phosphate group is released from pump protein.

- Pump protein is restored to the original conformation.

- K$^+$ ions are released into cell.

The cycle repeats, as the pump continues to move sodium and potassium ions while hydrolyzing ATP.

Co-transport is another active transport mechanism. The energy in an existing gradient of, for example, sodium ions is used to pump another ion or molecule up its concentration gradient. This is achieved by allowing the sodium ions to move down their concentration gradient at the same time as the other ion or molecule is moving up its gradient. Special membrane proteins couple these two processes.

Exocytosis and **endocytosis** are more complex forms of active transport. In exocytosis, a membrane-bound package (granule or vesicle) fuses its membrane with the outer membrane of the cell, allowing the material in the package to pass to the outside of the cell. In endocytosis, material on the outside of the cell is engulfed by an infolding of part of the cell membrane. As a consequence, a membrane-bound package, filled with material from outside of the cell, is formed inside the cell. Thus, roughly speaking, these two processes are the opposite of one another. Three forms of endocytosis are **phagocytosis**, **pinocytosis**, and **receptor-mediated endocytosis**.

Topic Test 3: Active Transport

True/False

1. Exocytosis releases material from cells.

2. Living cells have more sodium inside than outside.

3. The source of energy for active transport can be ATP hydrolysis.

Multiple Choice

4. The sodium-potassium pump (ATPase)
   a. pumps sodium from inside the cell to outside.
   b. pumps potassium from outside to inside.
   c. hydrolyzes ATP.
   d. is a protein.
   e. All of the above are correct.

5. Which of the following is NOT correct?
   a. Diffusion will result in a net flux of glucose from a 1 M solution to a 2 M solution.
   b. Active transport can pump molecules up their concentration gradients.
   c. The sodium-potassium ATPase can pump sodium ions out of a cell.

d. Concentration gradients across a membrane can generate osmotic pressure.

e. Cells will lose water in a hyperosmotic solution.

Topic Test 3: Answers

1. **True.** This is one way the cell eliminates waste materials. It also can be used by endocrine cells to release hormones and by nerve cells to release synaptic transmitter.

2. **False.** Consider sea water, with a high concentration of NaCl, which is "outside" of single-celled organisms in the ocean. Consider that all life started in the ocean, and perhaps this will help you to remember which ion is in high concentration outside of cells.

3. **True.** The hydrolysis of ATP often supplies the energy directly, or indirectly by having been used to establish other chemical gradients in the case of co-transport.

4. **e.** Each of the statements is correct.

5. **a.** Diffusion of glucose is in the opposite direction.

APPLICATION: DDT

Membranes are effective barriers to substances because of their hydrophobic nature. The phospholipid bilayer is especially important in blocking the passage of more hydrophilic materials across the membrane. However, the hydrophobic nature of membranes also can contribute to certain problems with other hydrophobic substances.

Humans have created a number of artificial chemical substances, which normally are rare or absent in nature. One example of such a substance is DDT, a pesticide that was used for years to control insects that destroy plant crops. Many populations of insects evolved resistance to DDT, and farmers began using it in even higher concentrations in an attempt to overcome the resistance.

DDT is hydrophobic, and also is long-lasting in the environment. Because it is hydrophobic, it is retained in membranes and other sites high in lipid concentration, such as fatty tissue. DDT became more concentrated at the top of food chains, as we will see in Chapter 23. Because of DDT, certain birds were threatened with extinction, as recognized by Rachel Carlson in her book *Silent Spring*. As a consequence, DDT use in the United States was banned.

Chapter Test

True/False

1. Both active and passive transport require energy.

2. Co-transport is an example of passive transport.

3. Facilitated diffusion is an example of active transport.

Multiple Choice

4. Two solutions are separated by a membrane, with 1 M KCl initially located in the solution on one side, but not on the other. At equilibrium,
   a. if only K^+ is permeable, it will have equal concentration on the two sides, while Cl^- will only be on one side.
   b. if both K^+ and Cl^- are permeable, they will remain on one side of the membrane.
   c. if only K^+ is permeable, almost all of it will move to the other side of the membrane, opposite Cl^-.
   d. if only H_2O is permeable, it will have a net flux toward the side without KCl present.
   e. if only K^+ is permeable, some of it will move to the other side, but the charge difference across the membrane will cause the rest to remain on the same side as the Cl^-.

5. An animal cell is placed in a hyperosmotic solution. The result is
   a. a shrinking of the cell.
   b. a swelling of the cell.
   c. no change in the size of the cell.
   d. a shrinking of the cell, followed by a re-swelling.
   e. no change at first, then a shrinking.

6. Concerning active transport,
   a. it can pump molecules up their concentration gradients.
   b. it uses energy.
   c. exocytosis is a form of active transport.
   d. endocytosis is a form of active transport.
   e. All of the above are correct.

7. Which of the following part(s) of a membrane is (are) nonpolar?
   a. The head portion of phospholipid molecules
   b. The part of proteins that protrudes out of the cell
   c. The carbohydrate part
   d. The tails of phospholipid molecules
   e. All of the above

Chapter Test Answers

1. **F** 2. **F** 3. **F** 4. **e** 5. **a** 6. **e** 7. **d**

Check Your Performance:

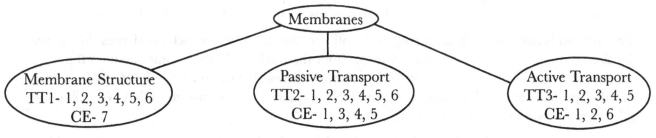

Key: TT = Topic Test: CE = Chapter Exam. Numbers indicate exam questions. Some questions are listed more than once if they refer to more than one topic.

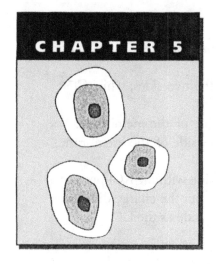

CHAPTER 5

Cellular Organization

Prokaryotic cells (bacteria and blue-green algae) do not have a membrane-bound nucleus, and only a few prokaryotic organisms have any internal membranes. Those membranes usually are for a single purpose, such as photosynthesis. In contrast, **eukaryotic** cells contain a nucleus. In fact, most eukaryotic cells contain many different internal membranes that organize and compartmentalize the activities of the cell. This chapter introduces some of the more important internal structures and organelles in eukaryotic cells.

ESSENTIAL BACKGROUND

- Membranes (Chapter 4)
- Proteins, phospholipids, and carbohydrates (Chapter 3)

TOPIC 1: NUCLEUS, ENDOPLASMIC RETICULUM, AND GOLGI APPARATUS

KEY POINTS

✓ *What does the nuclear envelope consist of?*

✓ *What is inside the nucleus?*

✓ *What is the role of endoplasmic reticulum in membrane biosynthesis?*

✓ *How are the different types of membranes sorted in cells?*

The **nucleus** of a eukaryotic cell is a large spherical body, bounded by a **nuclear envelope**. That nuclear envelope consists of two membranes—two lipid bilayers. Traffic into and out of the nucleus passes through specialized **pores** in the nuclear envelope. The pores are complex, protein-containing structures that are selective, allowing only certain materials to pass.

The nucleus is filled with **chromatin**, the "stuff" of chromosomes, which condenses during cell division, but otherwise is usually spread throughout the nucleus. Chromosomes contain the genetic material, DNA, structural proteins called **histones** that organize the DNA, and a variety of other proteins involved in duplicating the DNA, repairing it, and making RNA. We will consider chromosomes in more detail in Chapters 11 to 13.

A specialized region within the nucleus is called the **nucleolus**, where ribosomes are assembled. Assembled ribosomes leave the nucleus and are involved in protein synthesis in the cytosol.

We will return to the assembly of the ribosome, and its role in protein synthesis, in Chapter 10.

The nuclear envelope appears to be continuous with the **endoplasmic reticulum (ER)**. The ER is a maze of membrane-bound compartments. Two principal types of ER are **rough ER** and **smooth ER**. Rough ER is rough because of the presence of ribosomes bound to the surface of the ER membrane. In an electron micrograph, the ribosomes look like dots on the surface of the ER membrane. Proteins being made on the ribosomes attached to the rough ER are inserted into the interior (lumen) of the ER through special channels in the ER membrane. The inserted proteins include both membrane proteins and proteins that will be secreted from the cell (secretory proteins). They have been targeted to the rough ER by their first 30 to 40 amino acids, which serve as signal sequences. After the proteins begin to enter the ER, such signal sequences usually are removed. Carbohydrate groups are added to some of the proteins by enzymes found inside the ER lumen.

Membrane biosynthesis begins in the rough ER and continues in the smooth ER. The membrane and secretory proteins move through the rough ER lumen, and this appears to be continuous with the interior (lumen) of the smooth ER. The smooth ER appears smooth because it does not bind ribosomes, but that is not the only difference between rough and smooth ER. The smooth ER contains enzymes that catalyze the construction of membrane phospholipids, and also are involved in shifting such phospholipids from one side of the lipid bilayer to the other.

Notice that the ER, rough and smooth together, appears to be the site of the synthesis and assembly of membrane components—proteins, lipid, and carbohydrate. However, different membranes in the cell have different proteins and different lipid compositions as well. All of these proteins and lipids are intermixed in the membrane and lumen of the ER. Also in the lumen are secretory proteins, destined for export from the cell, as well as proteins that will be located in other organelles, such as the digestive enzymes destined for lysosomes. All of the different components of membranes and the lumen contents must be sorted into different packages. That sorting is done in the **Golgi apparatus**, a stack of membrane-bound sacs. **Transition vesicles** bud from the smooth ER, and carry the materials to one side of the Golgi apparatus. The Golgi sorts the materials, recycles some to the ER, and refines the carbohydrates attached to the proteins. Budding off from the opposite side of the Golgi are **vesicles**, **granules**, and **organelles**, each with its appropriate contents, ready to be shipped throughout the cell. These vesicles and granules serve a variety of roles in the cell, as we will see.

Topic Test 1: Nucleus, Endoplasmic Reticulum, and Golgi

True/False

1. Nuclear pores are nonselective.

2. The Golgi apparatus separates materials into membrane packages.

3. Chromosomes contain only DNA.

Multiple Choice

4. The Golgi apparatus
   a. is the primary site of biosynthesis of membrane lipids.
   b. is the primary site for the insertion of proteins from the cytoplasm into membranes.

c. is the site where the signal peptide is cut off of secretory proteins.

d. is the site of sorting of membrane proteins.

e. includes all of the above.

5. The nucleolus

a. is outside of the nucleus.

b. is the location of all of the DNA in a cell.

c. is the nucleus of a prokaryotic cell.

d. is the site of ribosome assembly.

e. includes all of the above.

Short Answer

6. Can nuclear pores be simple holes, or does the "door" to and from the nucleus need to be more complicated?

Topic Test 1: Answers

1. **False.** Nuclear pores are very selective, and even directional, allowing particular molecules in, and others out of, the nucleus.

2. **True.** This is the place in the cell where the sorting into different kinds of membrane "packages" takes place.

3. **False.** Chromosomes also contain a variety of proteins, from histones to enzymes.

4. **d.** All of the other events take place in the endoplasmic reticulum.

5. **d.** The nucleolus is inside the nucleus and contains the genes responsible for making pieces of RNA found in ribosomes, but does not contain all of the DNA of the nucleus.

6. If one considers the various kinds and sizes of molecules that must selectively pass the nuclear envelope unidirectionally, then it becomes clear that a simple hole or channel will not work. Large entities, such as RNA molecules, proteins, and even large multi-macromolecular complexes like ribosomes, pass through such pores. Recent work shows that the pore contains a highly complex system for properly transporting many of these large entities.

TOPIC 2: CELLULAR ORGANELLES

KEY POINTS

✓ *What are lysosomes and what do they do in cells?*

✓ *What do peroxisomes and glyoxysomes do?*

✓ *What are the roles of mitochondria and chloroplasts?*

✓ *What is the likely evolutionary origin of mitochondria and chloroplasts?*

A number of membrane-bound entities called **organelles** help to organize the activities of a living cell. Some of these have their origin at the Golgi apparatus, as described earlier, and others seem to be more independent. A typical cell will contain hundreds or thousands of organelles.

Typical animal cells contain several hundred **lysosomes**. These organelles contain digestive enzymes, and are acidic inside—pH 3 to 5. On their membrane surfaces, lysosomes have surface detectors that identify targets. They will fuse with the target and digest it. The digested building blocks diffuse back into the cytosol for reuse, while any residue can be discharged by exocytosis.

Mutations can produce inactive lysosomal enzymes, resulting in disease. One example is **Tay-Sachs disease**, which results from a damaged lipid-digesting enzyme. The undigested lipids accumulate, causing the lysosome to grow. As the lysosomes swell, they kill nerve cells, and death results.

The morphology (structure) of lysosomes varies, depending on what they have fused with. This is rather like the case of the morphology of your stomach, which changes with what you eat.

Peroxisomes carry oxidative enzymes, such as catalase, which breaks down hydrogen peroxide. Amino acids are oxidized in peroxisomes for use as an energy source by cells. Peroxisomes are not produced by the Golgi apparatus.

Glyoxysomes are found only in plants. They convert stored fats to carbohydrates, something that animals cannot do. Animals can convert carbohydrates to fat, as some of us realize all too well.

Chloroplasts and **mitochondria** are larger organelles that are involved in energy conversions. Chloroplasts are found in plants and some prokaryotes; mitochondria are found in most eukaryotic cells, including plants and animals. Mitochondria are the power stations of the cell. They convert the chemical bond energy in carbon-carbon and carbon-hydrogen bonds to the chemical bond energy in ATP. The ATP is used by the cell to do work. The fuel that the mitochondria use can be carbohydrate, fat, or protein. We will study that process of cellular respiration in great detail in Chapter 7. Chloroplasts convert light energy into chemical energy in the form of carbon-carbon and carbon-hydrogen bonds in carbohydrates, which are made by the chloroplasts from carbon dioxide and water. We will study that process in detail in Chapter 8.

There is considerable evidence favoring an unusual evolutionary origin for both mitochondria and chloroplasts. It is now thought that both started as independent prokaryotic organisms that were engulfed by other cells. Both organelles are surrounded by double membranes, rather like what would be expected if they originated as prokaryotic cells that were engulfed by other cells through a form of endocytosis (Chapter 4) called phagocytosis, which would add a second membrane layer. Each chloroplast and mitochondrion has a piece of DNA, which serves as further evidence that they once were independent organisms. The DNA codes for ribosomes, and they make their own ribosomes. Furthermore, those ribosomes resemble prokaryotic ribosomes rather than eukaryotic cell ribosomes.

Topic Test 2: Cellular Organelles

True/False

1. Glyoxysomes are organelles that convert fat to carbohydrate in animal cells.

2. Mitochondria are independent entities in eukaryotic cells, and do not require the eukaryotic cells to maintain themselves.

Multiple Choice

3. Lysosomes
   a. can have varying morphology.
   b. contain digestive enzymes.
   c. can cause death if they contain defective enzymes.
   d. are acidic inside.
   e. All of the above are correct.

4. Comparing chloroplasts and mitochondria,
   a. both are found in plants.
   b. mitochondria are in animals only; chloroplasts in plants only.
   c. both break down carbohydrates to carbon dioxide and water.
   d. both make carbohydrates from carbon dioxide and water.

Short Answer

5. Since animals cannot convert fats to carbohydrates, what can they do with fat?

Topic Test 2: Answers

1. **False.** They are found only in plant cells.

2. **False.** The cell provides many of the proteins that mitochondria require.

3. **e.** Each statement is correct.

4. **a.** Mitochondria are found in most cells of both plants and animals. An interesting exception is the red blood cells of humans, which carry oxygen for use in mitochondria throughout the body, but have no mitochondria of their own.

5. Animals can store the fat for insulation, for cushioning against injury, and for future energy needs. When energy is needed, fats serve as a rich source of energy for the production of ATP by mitochondria.

TOPIC 3: CYTOSKELETON AND CELL JUNCTIONS

KEY POINTS

✓ *What gives cells their shapes and allows them to move?*

✓ *What are the major filaments that are part of cytoskeleton?*

✓ *What linkages exist between neighboring cells in a tissue?*

Cytoskeleton

The **cytoskeleton** gives structural integrity to a cell, much as our skeletons help determine our shapes. The cytoskeleton consists of several different kinds of filamentous (long, stringlike) proteins that can be linked to the outer membrane of a cell. The protein filaments that make up the cytoskeleton not only serve as scaffolding, but also underlie motility (movement) and shape changes in cells.

Microfilaments are also known as **actin filaments** since they are composed of many linked actin proteins, plus associated proteins. Each actin monomer, or protein, is 43,000 daltons in size. Actin molecules link together in filaments to form a long, thin polymer that is helically intertwined and about 6 to 7 nanometers (each nanometer = 10^{-9} meters). Actin is found in all eukaryotic cells. Actin filaments in cells can be found close to the inside of the membrane, often linked to it through peripheral membrane proteins. Actin is known to play a role in maintenance of the shape of cells and changes in cell shape. It also is a principal component of muscle, and plays an important role in muscle contraction. Microfilaments also are involved in the flow of materials in plant cells, called **cytoplasmic streaming**.

A second part of the cytoskeleton, **microtubules** have the shape of hollow cylinders about 25 nanometers in diameter. Pairs of proteins, called alpha and beta **tubulin**, each about 55,000 to 57,000 daltons in size, make up this filament. Thirteen pairs of tubulin monomers make one turn of the cylinder, which can be very long. Microtubules, along with the microfilaments, play a role in cell shape and motility. Microtubules make up a major portion of the mitotic spindle apparatus, which is responsible for the movement of chromosomes during cell division. Microtubules are found in cilia and flagella in animals, and are involved in their wave motions. Cellular organelles can move along microtubules, propelled by special motor proteins that hydrolyze ATP as an energy source for the movement.

Both actin and tubulin can assemble, disassemble, and reassemble, thus serving as a changeable scaffolding for the cell. In contrast, the class of filaments known as **intermediate filaments** is more fixed. Intermediate filaments are about 10 nanometers in diameter, intermediate between microfilaments and microtubules (hence, the name). These are a class of protein-containing filaments, but the nature of the protein building blocks varies in different kinds of cells. In neurons, for example, intermediate filaments are called **neurofilaments** and contain three different kinds of proteins. Intermediate filaments are thought to give more permanent shape to cells.

Cell Junctions

Cells have specialized linkages to other cells. We consider three of these, all involving membrane proteins. **Tight junctions** can "fuse" two cells so tightly together that little material can pass between the cells. Tight junctions are specialized proteins that link from membrane to membrane. Tight junctions are found in skin, for example, helping us to maintain an inside environment separate from the outside one.

Desmosomes are more like "spot" welds that hold two cells tightly together at a point on their surfaces. The desmosome appears in electron micrographs to be a dense thickening of the linked membranes, with intermediate filaments present, extending into each cell from the thickening.

Unlike tight junctions and desmosomes, **gap junctions** allow for the passage of small molecules (about 1.5 nanometer and less) between cells. Gap junctions cause the formation of small channels between the two cells, with no contact to the outside. The cytoplasms of the two cells become interlinked electrically in such gap junctions, but molecules the size of most proteins and DNA cannot pass through the channels. Thus, gap junctions link cells and allow for rapid electrical signaling and passage of smaller molecules between the two cells.

Topic Test 3: Cytoskeleton and Cell Junctions

True/False

1. Actin filaments are 25 nanometers in diameter.

2. Actin proteins make microtubules.

3. Tight junctions allow the passage of materials between cells.

Multiple Choice

4. Which of the following cytoskeletal elements is a cylinder, about 25 nanometers in diameter?
   a. Microtubules
   b. Microfilaments
   c. Actin filaments
   d. Intermediate filaments
   e. Neurofilaments

5. Which of the following cell-cell contacts can prevent the passage of materials between cells (*outside of cells*), such as in the skin?
   a. Gap junctions
   b. Tight junctions
   c. Desmosomes
   d. Both a and b
   e. All of the above

6. Which component of cytoskeleton does not exhibit breakdown and resynthesis of its components into filaments?
   a. Intermediate filaments
   b. Microfilaments
   c. Microtubules
   d. Actin filaments
   e. Both b and d

Short Answer

7. Create a table about the cytoskeleton, showing the name of each filament, its diameter, and protein from which it is made.

Topic Test 3: Answers

1. **False.** Microtubules are 25 nanometers in diameter.

2. **False.** Actin proteins make microfilaments.

3. **False.** Gap junctions allow for the passage of small molecules and ions between cells.

4. **a.** Only microtubules are hollow cylinders.

5. **b.** Gap junctions and desmosomes alone will allow material to "leak" between cells.

6. **a.** Intermediate filaments are the most stable.

7.

Name	Diameter (nanometers)	Protein Monomer
Microfilament	6–7	Actin
Microtubule	25	Alpha and beta tubulin
Intermediate filament	10	Various

APPLICATION: VIRUS INFECTION OF CELLS

The wonderful machinery within cells carries out the functions of living organisms. Viruses are interesting entities on the edge of life that "borrow" cell machinery. Viruses alone cannot do all of the things that living cells can. While viruses are not alive, they do contain genetic material, generally surrounded by a protective coat of protein, and perhaps a membrane. Viruses must invade living cells to multiply. They are dormant until they infect a living cell. Once inside, their genetic material can be copied, and they multiply as their proteins are produced using the machinery of the living cell.

Viruses cause diseases ranging from colds and flu to herpes and AIDS. They are dangerous but fascinating infectious agents that have evolved to target specific types of cells. Mild viruses cause discomfort in their host, but other strains of virus can be life-threatening. Among the most lethal to humans are hemorrhagic fever viruses such as Ebola. The Ebola Zaire strain kills over 90% of infected individuals, generally within 2 weeks of infection. Infection spreads via body fluids, contact with which can be hard to avoid because such individuals begin to bleed throughout their bodies. In the United States, another hemorrhagic fever virus, hantavirus, has caused a number of deaths after spreading from rodents to humans. For more details on these deadly viruses, read *The Hot Zone* by Richard Preston. We will examine viruses in more detail in Chapter 14.

Chapter Test

True/False

1. Nuclear pores are simple holes in the nuclear envelope.

2. Microtubules are an integral part of desmosomes.

3. Chloroplasts create light energy from chemical energy.

Multiple Choice

4. The nucleus of a cell
   a. is surrounded by a double membrane called the nuclear envelope.
   b. contains RNA.
   c. contains DNA.
   d. contains proteins.
   e. All of the above are correct.

5. Which of the following is thought to have arisen in some eukaryotic organisms from an invading microorganism that became an essential component of these organisms?
   a. Lysosomes
   b. Cytoskeleton
   c. Golgi apparatus
   d. Chloroplasts
   e. Glyoxysomes

6. Which component of cytoskeleton also plays a role in muscle contraction?
   a. Microtubules
   b. Intermediate filaments
   c. Microfilaments
   d. Neurofilaments
   e. None of the above

Chapter Test Answers

1. **F** 2. **F** 3. **F** 4. **e** 5. **d** 6. **c**

Check Your Performance:

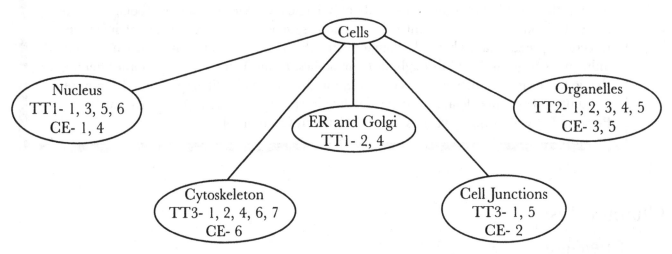

Key: TT = Topic Test; CE = Chapter Exam. Numbers indicate exam questions. Some questions are listed more than once if they refer to more than one topic.

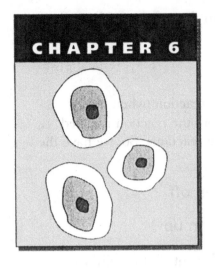

CHAPTER 6

Energy and Enzymes

In earlier chapters we learned about the physics and chemistry of atoms and molecules, and applied that knowledge to gain an understanding of the molecules of life, membranes, and cell organelles. Now it is time to learn how proteins work and how energy is used in cells. To gain that knowledge, we first return to fundamental physics and chemistry for another insight or two.

ESSENTIAL BACKGROUND

- **Equations**
- **Inequality signs (example "a > b" is read "a is greater than b")**

TOPIC 1: THERMODYNAMICS, HEAT, AND ENTROPY

KEY POINTS

✓ *What are the first and second laws of thermodynamics?*

✓ *What are energy, enthalpy, and entropy?*

✓ *What are endothermic and exothermic reactions?*

Energy comes in different forms, including mechanical, electrical, heat, chemical, and so on. These are interconvertible, but a couple of important physical laws describe limits on that interconvertibility.

The **first law of thermodynamics** states that energy can neither be created nor destroyed. Energy is conserved. Energy can change from one form to another, but the total amount of energy in the universe is a constant. If we take into account all the changes in energy during a process, there will be no change in total energy before and after the process. In other words, in a closed system (where all forms of energy are taken into account), $\Delta E = 0$ **[closed system]**. Living organisms are open systems, not closed ones. In living organisms we often are concerned with chemical reactions. In a typical chemical reaction an energy change takes place:

$$\Delta E = E_{products} - E_{reactants} \text{ [open system]}$$

In most reactions in living organisms, the energy difference between products and reactants is in the form of heat, and is related to the changes in chemical energy in the bonds of product molecules compared to reactant molecules. For instance, the chemical bonds in the products of a chemical reaction might have less energy than the bonds in the reactants, and the extra energy could be given off as heat. Such a reaction is called **exothermic**, and the amount of heat given off is expressed as follows:

$$\Delta H = H_{products} - H_{reactants}$$

where H stands for **enthalpy**, the heat of reaction. In an exothermic reaction (when heat is released), ΔH is negative, by definition. If heat is taken up in a reaction, the reaction is called **endothermic**—the products have a higher heat content than did the reactants. So, we have the following:

$$\Delta H < 0 \rightarrow \text{exothermic reaction} \rightarrow \text{heat given off}$$

$$\Delta H > 0 \rightarrow \text{endothermic reaction} \rightarrow \text{heat taken up}$$

A second factor influences the direction of chemical reactions, and of events in the universe more generally. This factor is **entropy (S)**. Entropy is a measure of randomness or disorder in a system. The greater the entropy, the greater the disorder. The second law of thermodynamics states that in a closed system, ΔS, the change in entropy, is greater than or equal to zero: $\Delta S \geq 0$ **[closed system]**. The universe gets more disordered with time.

Perhaps some examples of entropy would help. Consider a sand castle. Say you carefully build a sand castle, with towers and a moat, inside of a box. Now close the box and shake it up. The castle is gone, only a pile of sand remains. The pile of sand is more disordered, more random than the castle was. Close the box and shake it again . . . still no castle. The second law of thermodynamics says that order cannot arise out of disorder without help. Clearly a castle is extremely unlikely to arise spontaneously in this case. The second law really is a statistical one— there is an extremely low probability of the sand forming a castle again without outside help. One probably would have to keep shaking the box for longer than a lifetime.

In most common situations, such as a volume of gas, there are many more molecules than there are grains of sand in a sand castle, and the odds of a violation of the second law are even much lower, and not likely to happen during the entire history of the universe. Consider all of the air in the room you are in. What are the odds of all the air molecules being in one-half of the room, with a vacuum in the other half? If, instead, we somehow start with all of the molecules on one side, won't they quickly diffuse to fill the whole room? Diffusion is a good example of increasing entropy.

For another example of the direction of increasing entropy, see the web page. In the next section, we put these concepts of enthalpy and entropy together to understand the direction that chemical reactions proceed in.

Topic Test 1: Thermodynamics, Heat, and Entropy

True/False

1. Entropy is the heat of reaction.

2. The first law of thermodynamics states that energy is conserved in an open system.

3. An exothermic reaction releases heat.

Multiple Choice

4. Which of the following would necessarily violate the second law, but not necessarily the first law of thermodynamics?
   a. A living organism becomes more ordered with time.
   b. A newly discovered species generates its own energy, over many generations, with no need for any input or output of energy.
   c. All of the air molecules in a room move to one-half of the room by diffusion alone.
   d. An organism generates heat from chemical energy.

5. Which of the following would necessarily violate the first law of thermodynamics?
   a. A chemical reaction releases heat to the environment.
   b. Energy is converted from chemical form to mechanical form.
   c. In a closed system, there is a net loss of energy.
   d. In an open system, there is a gain of energy.
   e. All of the above would violate the first law of thermodynamics.

6. Two rooms, in contact with each other, constitute a closed system. They initially are at the same temperature. The temperature in one room decreases while the other increases, with no other changes occurring. This would necessarily
   a. violate the first law of thermodynamics.
   b. violate the second law of thermodynamics.
   c. not violate either the first or the second law.
   d. violate both the first law and the second law.

Short Answer

7. If, according to the second law, the universe is getting more disordered with time, how do living things get more ordered as they grow and reproduce?

Topic Test 1: Answers

1. **False.** Enthalpy is the heat of reaction.

2. **False.** Energy is conserved in a closed system.

3. **True.** ΔH is less than zero.

4. **c.** Answers a and d do not necessarily violate any physical law, and answer b would necessarily violate the first law.

5. **c.** The other answers would not necessarily violate any physical law.

6. **b.** The first law need not be violated.

7. Living things are open systems. They increase their own order by taking in energy and increasing disorder in their environments as heat is released.

TOPIC 2: FREE ENERGY AND COUPLING

KEY POINTS

✓ *What is free energy?*

✓ *What are exergonic and endergonic reactions?*

✓ *How does coupling allow otherwise endergonic reactions to proceed?*

Free energy is defined in terms of enthalpy and entropy. We are most interested in changes in free energy, ΔG. We define

$$\Delta G = \Delta H - T \cdot \Delta S,$$

where T is absolute temperature. The formula says that in a chemical reaction in living organisms (temperature and pressure constant), the free energy change is the change in enthalpy (heat of reaction) minus a correction for the change in disorder of the system.

The free energy change is most fundamental for any chemical reaction. If ΔG < 0, the reaction can proceed spontaneously, to form products from reactants. The reaction is called **exergonic**. It is spontaneous; free energy is given off as products are formed from reactants; and the products have less free energy than do the reactants. In contrast, when ΔG > 0, the reaction is called **endergonic** and will not occur spontaneously.

Many of the reactions that are needed by living organisms are endergonic. Examples include the making of macromolecules—proteins from amino acids, DNA and RNA from nucleotides, and glycogen from glucose. Free energy cannot increase spontaneously, so how do such reactions occur? They occur by coupling the endergonic reactions to other, exergonic reactions. If an endergonic reaction is coupled to an even more exergonic one, then the coupled pair of reactions can occur spontaneously.

We are quite familiar with coupled events of this sort because human-made machines do the same thing. Consider a car traveling up a hill. Without its engine running, a car moving up a hill is endergonic, but if one couples the burning of gasoline (an exergonic reaction to the movement of the car), one has an overall exergonic event.

Now consider the movement of sodium ions out of the cell by the sodium pump (Na-K ATPase) that we discussed in Chapter 4. That movement up the concentration gradient for sodium is endergonic. However, by coupling the movement of sodium ions to the hydrolysis of ATP, a very exergonic event occurs: The coupled events—sodium movement and ATP hydrolysis—have a net ΔG < 0, and the pump works spontaneously. In living organisms, ATP hydrolysis often is coupled to energy-requiring processes and reactions. Proteins, acting as enzymes, often are the coupling agents that link exergonic hydrolysis of ATP to endergonic events.

All living organisms are open systems, demanding free energy sources from their environments. Humans depend on the free energy in chemical bonds that we get from the food we eat. We get and stay organized at the expense of disorder in the rest of the universe. We do not violate the laws of thermodynamics any more than a car or air conditioner does!

In a chemical reaction, as products form from reactants, the value of ΔG changes, because ΔG is a function of the ratio of reactants to products. At equilibrium for a chemical reaction, ΔG = 0, and the rate of the forward reaction (forming products from reactants) is equal to the rate of the back-reaction (forming reactants from products).

Topic Test 2: Free Energy and Coupling

True/False

1. Exergonic reactions can be exothermic or endothermic.

2. Endergonic reactions can proceed spontaneously.

3. Living organisms get endergonic reactions to proceed by coupling them to exergonic reactions.

Multiple Choice

4. In a chemical reaction, at equilibrium,
   a. the forward rate of the reaction equals the back-reaction rate.
   b. pH = 7.
   c. $\Delta G < 0$.
   d. the second law of thermodynamics is violated.
   e. All of the above are correct.

5. A chemical reaction is exergonic and exothermic. The reaction
   a. can proceed spontaneously.
   b. will not occur spontaneously, but could if coupled to an endergonic reaction.
   c. has a $\Delta H > 0$.
   d. has a $\Delta G > 0$.

Short Answer

6. What is the ultimate source of free energy that living organisms depend on?

Topic Test 2: Answers

1. **True.** Notice the formula, above, defining ΔG. Even if ΔH is positive, so long as $(T \cdot \Delta S)$ is positive and of sufficient magnitude to offset ΔH, then ΔG will be negative, making the reaction exergonic.

2. **False.** Exergonic reactions can proceed spontaneously. Endergonic reactions only proceed when they are coupled to exergonic reactions.

3. **True.** So long as the summed ΔG for the two reactions is negative.

4. **a.** And $\Delta G = 0$.

5. **a.** ΔG and ΔH are < 0.

6. Life on earth ultimately depends on a coupling between the biosphere and sunlight. The free energy in sunlight is harnessed, with photosynthesis being the ultimate coupler. We will learn how photosynthesis works in Chapter 8.

TOPIC 3: HOW ENZYMES WORK

KEY POINTS

✓ *Why isn't spontaneous also instantaneous?*

✓ *What is activation energy?*

✓ *How do enzymes work? What do enzymes do to free energy and activation energy?*

All reactions with $\Delta G < 0$ can proceed spontaneously, so why haven't they all occurred already? Why hasn't the universe already run itself down to a state of disorder where there is nothing left but random fluctuations? Life wouldn't just be dull; it would not exist. The answer: Spontaneous is not necessarily fast. Spontaneous means without help, or self-generated. In fact, if all exergonic reactions proceeded quickly, most of our important macromolecules, from DNA to protein, would break apart, releasing free energy as they did.

Spontaneous is not instantaneous because of **activation energy**, the added energy that is needed for a reaction to occur. It is rather like a rock at the top of the hill needing a nudge to get started rolling down the hill. Activation energy is like that extra energy required to get things going. In a chemical reaction the activation energy is needed to begin to break existing chemical bonds so that new ones can be formed in their place. The required activation energy is "returned," with interest if a reaction is exergonic, but that fact does not help to get things going. The speed of a chemical reaction, the reaction rate, depends inversely on the size of the activation energy, E_A—the greater the activation energy, the slower the reaction proceeds. Much of what life is all about is increasing the reaction rates for the right set of reactions.

Enzymes are proteins that catalyze chemical reactions. Thus, enzymes are **catalysts** that speed the rate of a reaction without themselves being permanently changed (at the end of the reaction). Each enzyme speeds the rate of one chemical reaction, or a small number of closely related chemical reactions. A typical living cell contains thousands of different kinds of enzymes.

Enzymes speed reaction rates by lowering E_A. Enzymes **do not** influence the direction of a reaction—that is determined by ΔG. Enzymes **do not** change ΔG for a reaction, except as they change the ratio of products to reactants.

Enzymes bind to reactants and reduce E_A, allowing products to form more rapidly. It is not unusual for enzymes to produce a million-fold increase in reaction rate. The final ratio of product to reactant is determined by ΔG; enzymes act to speed both directions of the reaction.

Notice also that the size of ΔG does not influence the reaction rate. Consider the rock at the top of the hill. A rock is not more likely to start to roll if the hill happens to be twice as high. It just rolls farther once it starts.

Enzymes bind **substrates (reactants)** and convert the substrates to **products** at their **active sites**. The binding and catalytic activity at the active site is determined by the "fit" of the substrate to the enzyme and the nature of the chemical groups at the active site. The catalytic cycle of an enzyme consists of binding of the substrate(s) to the enzyme, conversion of substrate(s) to product(s) at the active site, and release of the product(s).

A number of factors influence the rate at which enzymes work. If the concentration of the reactants, or substrates, for the enzyme increases, then the reaction rate will also increase, up to a point. One can saturate the enzyme with substrate, and any further increase will not significantly increase the rate at which product is formed.

Because of the secondary and tertiary structure in proteins, and the chemical nature of the catalyzing part of the protein, both temperature and pH influence enzyme action. As temperature increases, so does enzyme activity, up to the point when the protein starts to unfold—secondary and tertiary structure are lost due to increased thermal energy, which changes the conformation of the protein (the enzyme denatures). Enzyme activity, which was increasing, starts to decline. This is one of the problems that occurs with very high fevers in humans.

Enzymes also have a pH where their activity is a maximum, and their activity declines if the pH is either raised or lowered from that maximum. In most cases, the pH is changing the ionic charges on the protein, and this results in a changing of secondary and tertiary structure, or an altering of charged groups at the active site.

Topic Test 3: How Enzymes Work

True/False

1. Spontaneous reactions occur rapidly.

2. The active site for an enzyme is that part of the enzyme that is involved in catalysis—the production of product from substrate.

3. Enzymes work by increasing E_A, the energy of activation.

Multiple Choice

4. Enzymes
   a. all have the same optimal temperature, at which the reaction rate is a maximum.
   b. all have the same optimal pH, at which the reaction rate is a maximum.
   c. increase the free energy released by a particular reaction.
   d. are catalysts that speed the rate of one reaction, or a few reactions.
   e. Both a and b are correct.

5. Which of the following can influence the rate at which an enzyme-catalyzed reaction takes place?
   a. pH
   b. Temperature
   c. Substrate (reactant) concentration
   d. Both a and c
   e. All of the above

Topic Test 3: Answers

1. **False.** Activation energy needs to be supplied before spontaneous reactions occur.

2. **True.** The active site consists of the shape and chemical nature of the amino acid side groups present, and usually depends on the conformation (secondary and tertiary structure) of the enzyme.

3. **False.** Enzymes work by decreasing E_A.

4. **d.** Different kinds of enzymes will have different "best" pH and temperature. For instance, the digestive enzymes we secrete into our stomachs work best in a more acidic environment. Also, enzymes are very selective in the reactions they catalyze.

5. **e.** All three factors influence reaction rates.

TOPIC 4: ENZYME REGULATION AND INHIBITION

KEY POINTS

✓ *What is feedback inhibition (negative feedback)?*

✓ *What is an allosteric site?*

✓ *What is the difference between competitive and noncompetitive inhibition?*

Many enzymes work in teams called **biochemical pathways**. Biochemical pathways convert an initial substrate to a final end-product through a series of enzyme-catalyzed reactions. There are intermediate products, each of which becomes the substrate for the next enzyme in the pathway:

$$V \xrightarrow{E_v} W \xrightarrow{E_w} X \xrightarrow{E_x} Y \xrightarrow{E_y} Z$$

where E_v to E_y are enzymes, V is the initial substrate, W to Y are intermediates, and Z is the final product. You will be studying such pathways and their consequences for as long as you study biology.

These pathways often exhibit special regulation. Some of the enzymes, typically the first enzyme in the pathway (E_v), have a second site, beyond their active site. The second site is called an **allosteric site**, and binding of a molecule at that site causes a change in the activity of the enzyme. This change is thought to be brought about by a change in the conformation (secondary or tertiary structure) of the protein. That is how a small molecule binding at the allosteric site can affect activity at a more distant active site. In a biochemical pathway, the binding usually causes a decrease in the activity of the enzyme, and the molecule that binds there is the end-product of the pathway (Z in the example). This is called **negative feedback**, or **feedback inhibition**, or end-product inhibition.

Notice that feedback inhibition is a means for controlling the amount of Z that is made. As the concentration of Z increases, more E_v will have Z allosterically bound to it, blocking further production of Z. If the amount of Z decreases, then there will be more active E_v enzymes, thus allowing production of Z. This kind of feedback inhibition is similar to the way that a thermostat works to regulate the temperature of a room.

Enzyme activity can be changed in other ways. Some enzymes have their activity regulated by the addition or subtraction of chemical groups, such as phosphate groups, to the enzyme.

There also are inhibitors of enzyme activity that act at the active site. Some of these are poisons, which, when made by plants, help to protect the plants against being eaten. **Competitive inhibitors** bind reversibly (temporarily) at the active site, or very near to it, so as to block the ability of the enzyme to bind substrate. They are called competitive inhibitors because they compete with substrate for the active site, and the relative concentrations of inhibitor and substrate determine the extent to which the enzyme is inhibited.

Noncompetitive inhibitors bind to other sites on the enzyme. Allosteric inhibitors are an example of noncompetitive inhibitors. Finally, **irreversible inhibitors** tend to form more permanent, covalent bonds with enzymes, blocking their function. Nerve gases and some insecticides are examples of irreversible inhibitors.

Topic Test 4: Enzyme Regulation and Inhibition

True/False

1. Allosteric inhibitors bind to the active site of an enzyme.

2. Noncompetitive inhibitors bind to the active site of an enzyme.

3. Competitive inhibitors bind to the active site of an enzyme.

Multiple Choice

4. Which of the following is NOT correct? Feedback inhibition (negative feedback)
   a. is a method of regulation of enzymes in biochemical pathways.
   b. can be produced by allosteric binding of a product to an enzyme that is found earlier in a biochemical pathway.
   c. generally is irreversible.
   d. can block the activity of an enzyme.
   e. generally is noncompetitive.

Short Answer

5. Distinguish between competitive and noncompetitive inhibition.

6. Distinguish between reversible and irreversible inhibition.

Topic Test 4: Answers

1. **False.** They bind to separate, allosteric sites.

2. **False.** They bind elsewhere on the enzyme and cause conformational changes affecting the active site.

3. **True.** They compete with the substrate.

4. **c.** Feedback inhibition needs to be reversible since the enzyme needs to produce product again when there is less end-product present. The allosteric binding of end-product is reversible. It generally is noncompetitive because the reversible binding is not to the active site.

5. Competitive inhibition is reversible and the inhibitor binds at, or very near to, the active site, thereby blocking access to the site by the substrate. Noncompetitive inhibitors bind to other sites, altering the conformation of the protein, and blocking enzyme action.

6. Reversible inhibitors bind temporarily to the enzyme. They can come on and off of the enzyme. Irreversible inhibitors bind more permanently to the enzyme, often by forming a covalent bond with the enzyme.

One example of an irreversible inhibitor of an enzyme is a class of compounds called **nerve gases**. These agents were developed for chemical warfare. Chemical warfare, including lethal mustard gases, was used during World War I. After that war, strong opposition to the use of chemical warfare agents arose. However, development of chemical warfare agents continued into the 1960s. It was during this period that modern nerve gases were produced and stockpiled by a number of nations, including the United States. Despite general opposition to chemical weapons, several nations have been accused of using them, and several years ago a radical group in Japan released nerve gas in a subway, killing and injuring a number of individuals.

Nerve gases act on the enzyme acetylcholinesterase (AChE). This enzyme is found on muscle cells and in some synapses in the brain, where it breaks down the neurotransmitter acetylcholine (see Chapter 21). Acetylcholine stimulates muscle contraction, and AChE terminates the stimulation. With nerve gas present, the acetylcholine stimulation of our skeletal muscles continues. Poisoned individuals undergo sweating, nausea, vomiting, and involuntary twitching and jerking, followed by confusion, drowsiness, coma, convulsions, and death from asphyxiation.

A variety of nerve gases have been made, including sarin, soman, VE, and VX. While called nerve gases because they are usually administered in an aerosol form, most actually are liquids. The reaction can be modeled as follows:

$$\underset{\underset{CH_3}{|}}{\overset{\overset{F}{|}}{R\!-\!O\!-\!P}}\!=\!O + AChE \rightarrow HF + \underset{\underset{CH_3}{|}}{\overset{\overset{AChE}{|}}{R\!-\!O\!-\!P}}\!=\!O \rightarrow dead$$

The covalent bond between AChE and a phosphorus-containing group in the nerve gas molecule blocks the active site of AChE, terminating enzyme action. An antidote that has been used is pralidoxime chloride, which reacts with the nerve gas, breaking the bond with AChE.

Some insecticides also work by blocking AChE. A number of insects are especially vulnerable to blockage of this enzyme, although widespread use of insecticides on crops and lawns is resulting in the development of resistant strains of the insects.

Interestingly, Ft. Detrick, Maryland, which once was a major research site in the United States for studying nerve gases, has now been converted into a research center for studying such dangerous viruses as Ebola and hantavirus (Chapter 5).

Chapter Test

True/False

1. Enzymes are proteins.

2. An enzyme does not change the ratio of product and substrate for a reaction at equilibrium.

3. Enzymes reduce the activation energy for a reaction.

4. Enzymes influence the rate of a chemical reaction, but not the direction.

5. A typical cell will have about 20 different kinds of enzymes present.

6. Irreversible inhibition is the same as noncompetitive inhibition.

Multiple Choice

7. Which of the following is correct?
    a. Energy can be neither created nor destroyed, but entropy can increase in a closed system.
    b. Exothermic reactions cannot occur spontaneously.
    c. Enthalpy, the heat of reaction, always is positive.
    d. Endothermic reactions cannot occur spontaneously.
    e. All of the above are correct.

8. Concerning the second law of thermodynamics,
    a. the universe is becoming more ordered.
    b. living things must violate the law because they become more ordered.
    c. the energy of the universe is being destroyed with time.
    d. the energy in a closed system increases as chemical reactions occur.
    e. none of the above are correct.

9. The free energy for a reaction is greater than zero. One can conclude that
    a. the reaction will proceed spontaneously.
    b. the reaction might occur if it is coupled to another reaction having a negative free energy change.
    c. the reaction will only occur in sunlight.
    d. all of the above are correct.
    e. none of the above are correct.

Short Answer

10. What do enzymes do to the free energy and activation energy of reactions that they catalyze?

Chapter Test Answers

1. **T** 2. **T** 3. **T** 4. **T** 5. **F** 6. **F** 7. **a** 8. **e** 9. **b**

10. Enzymes do not change free energy and decrease activation energy.

Check Your Performance:

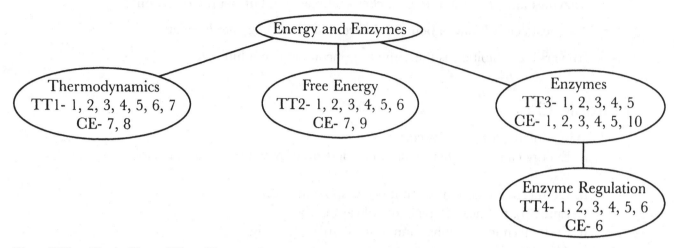

Key: TT = Topic Test; CE = Chapter Exam. Numbers indicate exam questions. Some questions are listed more than once if they refer to more than one topic.

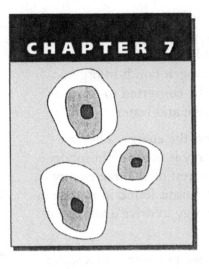

CHAPTER 7

Energy Metabolism and Cell Respiration

Life requires energy. That energy is used for making new molecules that are used in growth, maintenance, and reproduction; in movement; for pumping ions and molecules up concentration gradients; and in other ways. Most organisms extract the energy that they need from organic compounds. Even plants, after harnessing energy from sunlight and storing it as chemical energy in organic molecules, need a mechanism for extracting the energy. Plants and animals use cellular respiration, which extracts some of the chemical energy in organic molecules. During cell respiration, the energy is used to make ATP. In eukaryotic cells, much of that ATP production occurs in mitochondria.

ESSENTIAL BACKGROUND

- ATP, ADP (Chapter 3)
- ΔG (Chapter 6)
- Carbohydrates (Chapter 3)

TOPIC 1: OXIDATION, REDUCTION, AND NADH

KEY POINTS

✓ *What is the overall reaction for releasing free energy from carbohydrates, such as glucose?*

✓ *What are oxidation and reduction?*

✓ *What are NAD^+ and NADH?*

The general process of **cellular respiration** can be given as follows:

Organic compounds + oxygen → carbon dioxide + water + energy

We will use glucose as a model organic compound, and later in the chapter describe how other foods, such as fats, proteins, and other carbohydrates, feed into the same process. For glucose the formula is as follows:

$$C_6H_{12}O_6 + 6\ O_2 \rightarrow 6\ CO_2 + 6\ H_2O; \quad \Delta G° = -686\ kcal/mole$$

(Note: $\Delta G°$ is ΔG when the concentrations of reactants and products are equal.)

The key to understanding chemical energy in organic molecules is to realize that there are different amounts of chemical energy in different kinds of chemical bonds. In particular, bonds between carbon atoms (carbon-carbon bonds) and bonds between carbon and hydrogen (carbon-hydrogen bonds) contain more free energy than do bonds between oxygen and carbon, or between oxygen and hydrogen. Oxygen is quite **electronegative**—it holds on to electrons more tightly than does carbon or hydrogen. Carbon-carbon and carbon-hydrogen bonds in organic molecules can be viewed as sources of free energy if those bonds can be converted to oxygen-carbon and oxygen-hydrogen bonds, such as are found in carbon dioxide and water.

We could just burn the glucose and release heat, but living cells harness the energy. They are able to control the reaction so that about half of the released free energy is used to phosphorylate (add P_i—inorganic phosphate, Chapter 3) ADP, forming ATP. The rest of the free energy is released as heat. Making most of the ATP requires some special mechanisms found in mitochondria. In living organisms, ATP is immediately usable as a source of energy to drive a variety of otherwise endergonic processes.

The chemical reactions involve a transfer of electrons from one molecule to another in a series of reactions that gradually extract the energy. We speak of **oxidation-reduction reactions**, or **redox** reactions for short. Oxidation is defined as a loss of electrons; reduction is a gain of electrons. Notice the unusual "reduction is a gain" and perhaps that will help you to remember which is which. You also will need to understand what is meant by a loss and a gain. Most easily, whole electrons can be lost or gained by molecules, but the terms oxidation and reduction also apply when there is merely a change in the degree of sharing of electrons. Thus, in our overall model reaction for glucose, $C_6H_{12}O_6 + 6\ O_2 \rightarrow 6\ CO_2 + 6\ H_2O$, the glucose molecule is said to be oxidized and is called a **reducing agent**. Oxygen is reduced; it is the oxidizing agent. Notice that the oxygen molecules on the left side of the equation were sharing electrons equally. During the reaction, they gain electrons, or get reduced, because they become linked to carbon or hydrogen atoms and bind the electrons more tightly than do carbon and hydrogen. The electrons in the bond spend more time around the oxygen atom, so it has gained electrons, and thereby has undergone reduction, or has been reduced. Always, when something gets oxidized, something else gets reduced.

We will see that hydrogen atoms, with their electrons, get stripped from the glucose as it is processed, but they do not go directly to oxygen. The electrons and their associated free energy usually pass through an intermediate, **NADH**. For example, a dehydrogenase enzyme will pass a pair of electrons from part of the glucose molecule as follows:

$$\mathrm{H-\overset{\displaystyle |}{\underset{\displaystyle |}{C}}-OH + NAD^+ \rightarrow\ -\overset{\displaystyle |}{C}\!=\!O + NADH + H^+}$$

Two electrons plus two protons are released from one part of the glucose molecule and both electrons plus one of the protons combine with NAD^+ to form NADH. NAD^+ is an oxidizing agent; it is reduced. Each NADH represents stored energy. The oxidation of NADH back to NAD^+ releases energy:

$$\mathrm{NADH + H^+ + \tfrac{1}{2}\,O_2 \rightarrow \rightarrow \rightarrow NAD^+ + H_2O; \quad \Delta G^\circ = -53\,kcal/mole}$$

The arrows indicate that a series of steps is involved in the release of energy. We examine those steps later in this chapter.

Topic Test 1: Oxidation, Reduction, and NADH

True/False

1. If a molecule is oxidized, it gains electrons.

2. A reducing agent is oxidized.

3. An oxidizing agent is reduced.

Multiple Choice

For questions 4 through 7, consider the following reaction and options:

Reaction: $CH_4 + 2\,O_2 \rightarrow CO_2 + 2\,H_2O$

Options:
- a. CH_4
- b. O_2
- c. Only the hydrogens in CH_4
- d. Both CH_4 and O_2
- e. None of the above

4. Which is a reducing agent?

5. Which is an oxidizing agent?

6. Which is reduced?

7. Which is oxidized?

8. Which of the following bonds has more free energy?
   - a. Carbon-hydrogen
   - b. Carbon-oxygen
   - c. Hydrogen-oxygen
   - d. All the above have the same amount of free energy.

Short Answer

9. Given that $\Delta G = 8\text{--}10\,\text{kcal/mole}$ for the formation of ATP from ADP (it depends on the ration of ATP to ADP in a cell), and given the ΔG provided above for NADH oxidation, about how many ATPs could be formed from oxidation of one NADH (assume about a 50% efficiency).

Topic Test 1: Answers

1. **False.** It loses electrons.

2. **True.** A reducing agent reduces another compound. In doing so, it gets oxidized—it must lose electrons.

3. **True.** Same reasoning as in question 2.

4. **a.** Both carbon and hydrogen in CH_4 lose (partial) electrons because both kinds of atoms end up bonded to oxygen.

5. **b.** Oxygen often is an oxidizing agent: It gains electrons, and so whatever it combines with must lose electrons, or be oxidized.

6. **b.** The oxidizing agent is reduced.

7. **a.** The reducing agent is oxidized.

8. **a.** The extra energy in carbon-hydrogen and carbon-carbon bonds is harnessed during cell respiration.

9. The ratio is about 53/9, which is about 6. If efficiency is 50%, the number of ATPs formed per NADH would be about 3. This is about what has been observed.

TOPIC 2: GLYCOLYSIS

KEY POINTS

✓ *What happens during glycolysis?*

✓ *Where does glycolysis take place?*

✓ *How much ATP is produced and what is the end-product of glycolysis?*

There are three stages to cell respiration: glycolysis, Krebs cycle, and electron transport and oxidative phosphorylation. **Glycolysis** takes place in the cytosol of the cell; the remaining two take place in mitochondria. Glycolysis begins the breakdown of glucose. There is a net conversion of a little energy into ATP during glycolysis. This is achieved by **substrate-level phosphorylation**; that is, an enzyme directly catalyzes the transfer of a phosphate group from a substrate molecule to ADP, forming ATP.

There is a chain of 10 enzymatic steps during glycolysis. This chain of reactions converts each 6-carbon glucose molecule to two 3-carbon pyruvate molecules. A net of two ATP molecules and two NADH molecules also is formed for each glucose. Actually two ATP molecules are hydrolyzed and four are made, making a net of two ATP molecules. The overall reaction can be simply represented as follows:

$$\text{Glucose} + 2 \text{ ADP} + 2 \text{ P}_i + 2 \text{ NAD}^+ \rightarrow 2 \text{ pyruvate} + 2 \text{ ATP} + 2 \text{ NADH} + 2 \text{ H}^+$$

No carbon dioxide has been released yet, and no oxygen has been used yet. Glycolysis releases less than one-fourth of the energy that will be obtained from breaking down each glucose molecule fully to carbon dioxide and water. One of the early enzymes in the glycolysis pathway is called **phosphofructokinase**, which is an allosteric enzyme that serves as a control point for cell respiration, as we will see later.

The end-product of glycolysis, **pyruvate**, can have one of two fates. If oxygen is insufficient, fermentation occurs, a process we consider later in this chapter. When there is adequate oxygen present, pyruvate is modified for entry into the Krebs cycle as acetyl-coenzyme A (acetyl-CoA), a carbon dioxide is released from the pyruvate molecule, and NADH is formed:

$$\text{Pyruvate} + \text{NAD}^+ + \text{CoA} \rightarrow \text{acetyl-CoA} + CO_2 + \text{NADH} + \text{H}^+$$

The **acetyl-CoA** is the entry molecule for the Krebs cycle, which is considered in the next section. The acetyl group is the two-carbon fragment that remains after carbon dioxide has been removed from the pyruvate. Remember that there are two pyruvates, and thus two acetyl groups, generated from each glucose.

Topic Test 2: Glycolysis

True/False

1. Glycolysis takes place in mitochondria.

2. No oxygen is required during glycolysis.

3. During glycolysis, 2 pyruvate, 2 ATP, and 2 NADH molecules are formed per glucose.

Multiple Choice

4. Substrate-level phosphorylation
   a. involves the addition of a phosphate group to ADP.
   b. is catalyzed by enzymes.
   c. involves the removal of a phosphate group from a substrate.
   d. All of the above are correct.

Short Answer

5. Per glucose molecule, how many acetyl groups are formed for entry into the Krebs cycle?

Topic Test 2: Answers

1. **False.** Glycolysis takes place outside of mitochondria, in the cytosol, but the next two stages, Krebs cycle and electron transport and oxidative phosphorylation, do occur in mitochondria.

2. **True.** This becomes an important consideration for fermentation.

3. **True.**

4. **d.** Substrate-level phosphorylation is one of two ways that ATP is made during cell respiration.

5. Two acetyl groups are formed, one from each pyruvate.

TOPIC 3: KREBS (CITRIC ACID) CYCLE

KEY POINTS

✓ *What does the Krebs cycle consist of?*

✓ *What are the inputs and outputs of the Krebs cycle?*

The second stage in cell respiration is the **Krebs cycle**, which also is known as the **citric acid cycle**, or **TCA cycle**. Hans Krebs is the scientist largely responsible for describing the cycle. The cycle begins as the acetyl group (two carbons) from acetyl-CoA is linked to **oxaloacetate** (four carbons) to form **citrate**, a six-carbon compound. This is the first of eight steps, each catalyzed by a different enzyme, that actually end with the formation of oxaloacetate, bringing us back to the starting point, ready to accept another acetyl group. This series of steps is called a cycle because the pathway closes back around to form a ring of reactions. The enzymes in this cycle are located inside of mitochondria, in the mitochondrial matrix.

If you consider such cycles of reactions for a moment, you will begin to realize that the inputs and outputs to such a cycle become most important. What is consumed and what is produced each time the cycle is run? As the six-carbon chain in citrate is reduced to a four-carbon chain in oxaloacetate, two carbon dioxide molecules are removed. In addition, a molecule of ATP is produced by substrate-level phosphorylation. Three NADH molecules are formed from NAD$^+$, and one **FADH$_2$**, a molecule similar to NADH, is formed as well. All of this output is from one acetyl group entering the cycle. To account for one glucose molecule, which contributes two acetyl groups, we have to double these numbers.

Thus far in the overall process of cell respiration, we have the following:

Stage	Input	Output
Glycolysis	1 Glucose	2 ATP + 2 NADH + 2 pyruvate
Pyruvate to acetyl-CoA	2 Pyruvate	2 NADH + 2 acetyl + 2 CO$_2$
Krebs cycle	2 Acetyl	2 ATP + 6 NADH + 2 FADH$_2$ + 4 CO$_2$ (+6 H$^+$)

Notice that we have made a total of six carbon dioxide molecules, accounting for the six carbon atoms in glucose. We have, to this point, made six ATP molecules from ADP + P$_i$ by substrate-level phosphorylation, but we have a total of 12 (NADH + FADH$_2$) molecules, one for each of the carbon-carbon and carbon-hydrogen bonds in glucose. Notice also that, to this point, although we have made the carbon dioxide, we have not used oxygen. We will use the oxygen in the next stage of the process, and will be generating water with it, using the hydrogen molecules attached to NADH and FADH$_2$. In fact, that stage will generate enough water to offset the six molecules that were used up to this point (not shown above) plus six more, which will finally bring into balance the equation for the overall reaction that we gave at the beginning of the chapter:

$$C_6H_{12}O_6 + 6\ O_2 \rightarrow 6\ CO_2 + 6\ H_2O$$

Topic Test 3: Krebs Cycle

True/False

1. Per molecule of glucose, the Krebs cycle generates six carbon dioxide molecules.

2. The Krebs cycle begins with glucose.

3. The number of carbon-carbon plus carbon-hydrogen bonds in glucose is equal to the number of molecules of NADH and FADH$_2$ that are made during the glycolysis/Krebs cycle reactions.

Multiple Choice

4. During the Krebs cycle
   a. ATP is formed by substrate-level phosphorylation.
   b. NADH is formed.
   c. FADH$_2$ is formed.

d. CO_2 is formed.

e. All of the above are correct.

Topic Test 3: Answers

1. **False.** Four are generated during the Krebs cycle, and two more during the conversion of pyruvate to acetyl-CoA.

2. **False.** The Krebs cycle begins with acetyl groups, which were generated by glycolysis plus the conversion of pyruvate to acetyl groups.

3. **True.** There are 12 such bonds and 12 such high-energy products. If one examines the detailed steps in glycolysis and Krebs cycle (which we have not done here), it can be seen that each time one of the carbon-carbon or carbon-hydrogen bonds is lost, one of the NADH or $FADH_2$ molecules is formed from NAD^+ or FAD.

4. **e.** All of these are products of the Krebs cycle.

TOPIC 4: ELECTRON TRANSPORT CHAIN AND OXIDATIVE PHOSPHORYLATION

KEY POINTS

✓ *What does the electron transport chain do?*

✓ *How is the energy from electron transport stored?*

✓ *Where do the electrons end up?*

✓ *What does ATP synthase do?*

The final stage in cell respiration begins with the donation of two electrons from each NADH and $FADH_2$ to an **electron transport (respiratory) chain** found in the inner membrane of mitochondria (remember that mitochondria have two membranes—see Chapter 5). The NADH (and $FADH_2$) is oxidized (lose electrons) as the first member of the chain, NADH-Q reductase (a specialized protein), is reduced (gains the electrons).

This oxidation-reduction continues down the chain, which consists of a number of specialized molecules, mostly proteins and many containing iron, which actually serve as temporary acceptors for the electrons. Each member of the chain is reduced as it gets electrons and is oxidized as it passes them on to the next member. $FADH_2$ electrons follow a path similar to NADH electrons, but enter a little later in the chain because $FADH_2$ has a bit less free energy content in its electron pair than does NADH.

As the electrons travel along the chain, they gradually get more tightly bound, thereby losing free energy. That free energy is used to pump hydrogen ions (protons) from the inside of the mitochondrion to the space between the two mitochondrial membranes. Thus, a **proton (hydrogen ion) gradient** builds across the inner membrane, with the intermembrane space becoming more acidic (higher concentration of H^+).

At the bottom of the electron transport chain, the electrons are donated to oxygen, which together with hydrogen ions forms water. For every two NADH + H^+ molecules, one oxygen molecule (O_2) is converted to two water molecules.

The energy now is stored in the form of a proton gradient. Special channels allow the pumped hydrogen ions to return to the inside of the mitochondrion (the inside is called the matrix). These special channels are part of the protein **ATP synthase**. As the protons pass through ATP synthase, ATP is made from ADP + P_i. Thus, the hydrogen ion gradient is a useful intermediate, storing the free energy that was released as the electrons pass down the electron transport chain. That energy, sometimes called a proton-motive force, is used to form ATP by ATP synthase as the protons return to the matrix of the mitochondrion.

Chemiosmosis is the term used to describe the use of a proton gradient to couple the oxidation-reduction reactions in electron transport to the production of ATP by ATP synthase. Chloroplasts use the same mechanism of chemiosmosis to generate ATP during photosynthesis, as we will see in the next chapter. **Oxidative phosphorylation** refers to the linking of the (oxidative) flow of electrons along the respiratory chain to the phosphorylation of ADP to form ATP.

During this final stage of cell respiration, a net of about 32 ATP molecules is formed by oxidative phosphorylation from the 12 NADH + $FADH_2$ molecules generated by oxidation of 1 molecule of glucose. Remembering that we made 2 ATP molecules by substrate-level phosphorylation during glycolysis, and that another 2 were similarly made during the Krebs cycle, a total of about 36 ATP molecules are produced per glucose molecule during cell respiration. That production of ATP is the main output of the process, and the ATP is used by living cells for doing a variety of different kinds of work.

 For further information on catabolism and its regulation, see the web page.

Topic Test 4: Electron Transport Chain and Oxidative Phosphorylation

True/False

1. Free energy is made available as electrons travel along the electron transport chain.

2. ATP is generated directly by the energy released as electrons flow along the electron transport chain.

3. During oxidative phosphorylation, a proton gradient is used to form ATP.

4. Most of the members of the electron transport chain are specialized proteins.

Multiple Choice

5. When is most ATP formed during (aerobic) cell respiration?
   a. During glycolysis
   b. During Krebs cycle
   c. During electron transport and oxidative phosphorylation
   d. By substrate-level phosphorylation
   e. Both b and d

6. What directly donates most of the electrons to the electron transport chain in mitochondria?
   a. ATP
   b. NAD^+

        c. ADP

        d. NADH

        e. H_2O

    7. During which of the following processes does substrate-level phosphorylation occur?

        a. Glycolysis

        b. Krebs cycle

        c. Electron transport chain

        d. Pyruvate-to-acetyl-CoA conversion

        e. Both a and b

    8. During which of the following processes is NADH generated from NAD^+?

        a. Glycolysis

        b. Krebs cycle

        c. Electron transport chain

        d. Pyruvate-to-acetyl-CoA conversion

        e. a, b, and d

Topic Test 4: Answers

    1. **True.** That free energy is used to pump protons across the inner mitochondrial membrane, and ultimately is used to produce ATP.

    2. **False.** The ATP is generated by ATP synthase, using the proton gradient.

    3. **True.** ATP synthase uses the gradient to produce the ATP.

    4. **True.** Most are proteins that contain iron. Some contain heme groups, the same kind of group as is found in hemoglobin, which is used to transport oxygen in blood.

    5. **c.** About 32 of the 36 ATP molecules formed per glucose are formed by oxidative phosphorylation. However, notice that the NADH that provides the free energy source for electron transport was made during glycolysis, during the pyruvate-to-acetyl-CoA conversion, and during the Krebs cycle.

    6. **d.** NADH donates them as it is converted back to NAD^+.

    7. **e.** There is substrate-level phosphorylation during both indicated steps.

    8. **e.** NADH is generated during the three indicated steps.

TOPIC 5: FERMENTATION

KEY POINTS

    ✓ *What happens when there is insufficient oxygen available to run normal cell respiration?*

    ✓ *Describe two kinds of fermentation.*

    ✓ *What are strict aerobes, facultative anaerobes, and strict anaerobes?*

With oxygen present (aerobic conditions), cell respiration proceeds. Without oxygen, oxidative phosphorylation comes to a halt—there is no oxygen to accept electrons at the end of the electron transport chain. This situation leaves glycolysis as a source of ATP. Remember that two ATP molecules are generated per glucose molecule during glycolysis. However, there is a problem with using glycolysis alone as a continuing source of ATP. During glycolysis, NAD⁺ is reduced to NADH. Without some means of oxidizing NADH back to NAD⁺, we would quickly run out of NAD⁺, and glycolysis would halt.

Two different "fixes" for this dilemma are found in nature, and both are referred to as **fermentation**. Fermentation generates ATP by substrate-level phosphorylation in glycolysis, and nicely solves the problem of regenerating NAD⁺ by using pyruvate, the normal product of glycolysis.

In humans, when we go into oxygen debt during heavy exercise, we generate lactic acid from the pyruvate:

$$\text{Pyruvate} + \text{NADH} + \text{H}^+ \rightarrow \text{lactate} + \text{NAD}^+$$

The pyruvate is reduced to **lactate** by NADH, regenerating NAD⁺. This allows us to make some additional ATP, but we eventually slow down as lactic acid builds up in our muscles and blood. The same process of lactic acid fermentation is the general method of energy production by some fungi and bacteria that we use for making cheese and yogurt.

A second form of fermentation uses pyruvate and NADH to produce **ethanol** (ethyl alcohol) in two steps:

$$\text{Pyruvate} \rightarrow \text{acetaldehyde}$$

$$\text{Acetaldehyde} + \text{NADH} + \text{H}^+ \rightarrow \text{ethanol} + \text{NAD}^+$$

Yeast and some bacteria carry out alcohol fermentation under anaerobic (no oxygen) conditions. We use yeast to generate wine and beer.

Pyruvate represents a fork in the road. If oxygen levels are adequate, pyruvate continues through cell respiration and more ATP is produced as oxygen is consumed. If oxygen levels are low, fermentation occurs, allowing for the continuing production of a little ATP along with either lactic acid or ethanol as a by-product.

Strict aerobes can survive only in an atmosphere that contains oxygen. Both plants and animals are strict aerobes. **Facultative anaerobes**, such as yeast and most bacteria, can exist with or without oxygen, and can make ATP by either fermentation or aerobic respiration. A few organisms are **strict anaerobes**, which can make ATP and survive only in oxygen-free conditions. They tend to live deep in soil, in stagnant ponds, or in cans of improperly sterilized food.

Topic Test 5: Fermentation

True/False

1. Fermentation can generate the kind of alcohol that we drink in beer and wine.

2. Fermentation can generate lactic acid, which interferes with muscle activity.

3. Facultative anaerobes can survive with or without oxygen.

Multiple Choice

4. Concerning fermentation,
   a. in humans, fermentation produces ethanol from pyruvate.
   b. in yeast, fermentation does not occur.

c. in humans, fermentation cannot occur.

d. in yeast, fermentation occurs only when excess oxygen is present.

e. All the above are false.

5. Which of the statements below is false? In both fermentation and cell respiration,

a. pyruvate is an intermediate.

b. the reactions of glycolysis proceed.

c. ATP is produced.

d. electron transport continues to occur.

e. NADH cycles back to NAD$^+$.

Topic Test 5: Answers

1. **True.** Yeast is able to carry out these reactions.

2. **True.** Humans generate lactate, which reduces the effectiveness of muscle activity.

3. **True.**

4. **e.** Each of the other statements is incorrect.

5. **d.** Electron transport only occurs during cell respiration as it requires oxygen.

APPLICATION: MITOCHONDRIA AND AGING

During evolution, the introduction of mitochondria into eukaryotic cells became a kind of a "pact with the devil." Eukaryotic cells are able to extract more useable free energy from food because of the Krebs cycle and electron transport occurring in mitochondria, but in return the invading organism that became the mitochondrion is an essential part of all eukaryotic organisms. There are hundreds of trillions of mitochondria in each human, for example.

Even more significantly, mitochondria require oxygen to extract the additional energy from food. Oxygen itself is an oxidizing agent, and in mitochondria it can be converted into a very damaging entity. Occasionally oxygen gains an electron from the electron transport chain and then gets loose rather than completing its conversion to water. That oxygen with an unpaired extra electron is a form of **free radical** and can cause considerable oxidative damage to our cells.

Free radicals are very reactive and can donate their unpaired electron to other molecules, such as proteins, lipids, or DNA. The recipient molecule is modified, and the free radical can be passed to other molecules as reactions continue to occur, producing a chain of altered molecules for each oxygen free radical.

Humans have antioxidants that quench free radicals, and have repair systems that eliminate most of the molecules damaged by free radicals. Such protection is not perfect. Even though only a small fraction of free radical–damaged molecules go unrepaired, it is thought that the accumulation of such damaged molecules contributes to aging. While all of the underlying causes of aging are not fully understood, evidence is mounting that free radical damage is one of the more important.

Chapter Test

True/False

1. During glycolysis, a net of two molecules of ATP is generated per molecule of glucose.

2. During Krebs cycle, two molecules of ATP are generated per molecule of glucose.

3. During cell respiration, six molecules of carbon dioxide are generated per molecule of glucose.

Multiple Choice

4. A human muscle cell is making significant quantities of lactate (lactic acid). Which of the following is correct?
   a. The proton gradient across the inner mitochondrial membrane must be continuing to increase.
   b. Pyruvate concentrations in the cell must be continuing to increase.
   c. There probably is a lack of sufficient oxygen in the cell.
   d. The muscle cell no longer is making ATP.
   e. All of the above are correct.

5. ATP synthase
   a. makes ATP from ADP (+ P_i) using energy from a proton gradient.
   b. makes ATP by substrate-level phosphorylation.
   c. makes ATP directly from NADH as a substrate.
   d. Both b and c are correct.

6. In fermentation, the net production of ATP per glucose molecule is
   a. about 32 to 35 molecules.
   b. exactly 36 molecules.
   c. 12 molecules, one for each carbon-carbon or carbon-hydrogen bond in glucose.
   d. 10 molecules.
   e. 2 molecules.

Chapter Test Answers

1. **T** 2. **T** 3. **T** 4. **c** 5. **a** 6. **e**

Check Your Performance:

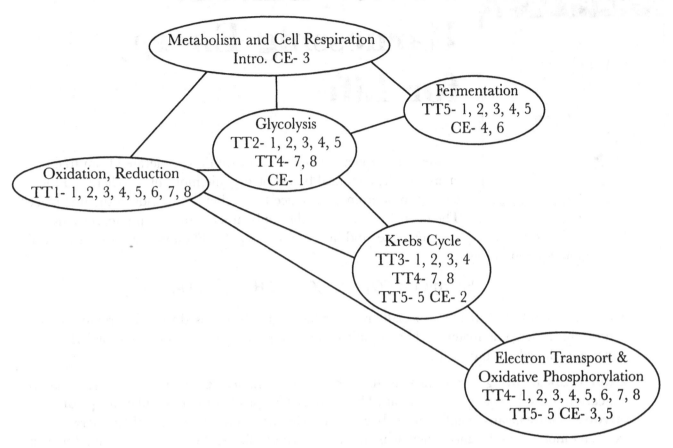

Metabolism and Cell Respiration
Intro. CE- 3

Fermentation
TT5- 1, 2, 3, 4, 5
CE- 4, 6

Glycolysis
TT2- 1, 2, 3, 4, 5
TT4- 7, 8
CE- 1

Oxidation, Reduction
TT1- 1, 2, 3, 4, 5, 6, 7, 8

Krebs Cycle
TT3- 1, 2, 3, 4
TT4- 7, 8
TT5- 5 CE- 2

Electron Transport &
Oxidative Phosphorylation
TT4- 1, 2, 3, 4, 5, 6, 7, 8
TT5- 5 CE- 3, 5

Key: TT = Topic Test; CE = Chapter Exam. Numbers indicate exam questions. Some questions are listed more than once if they refer to more than one topic.

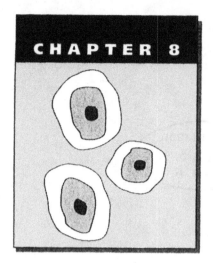

CHAPTER 8

Photosynthesis: Harnessing Energy for Life

In the last chapter, we explored how ATP is generated from free energy in the chemical bonds of organic molecules. Where did the energy come from to produce those chemical bonds in the first place? The energy comes from the sunlight, captured by photosynthesis.

Photosynthesis occurs in plants, in algae, and in some prokaryotes. Photosynthesis turns the cell respiration equation around:

$$\text{Energy} + 6\ CO_2 + 6\ H_2O \rightarrow C_6H_{12}O_6 + 6\ O_2,$$

where energy is from sunlight, especially red and blue light, and is used to build organic compounds from inorganic material. The organic compounds, in turn, nourish almost all living organisms, directly or indirectly.

Autotrophs are living organisms that make their organic molecules from inorganic raw materials: carbon dioxide, water, and minerals. Plants are autotrophs, and in particular are **photoautotrophs** (*photo* means "light"). Some bacteria are chemoautotrophs and get their energy from oxidizing inorganic substances such as methane or certain sulfur-containing molecules. The rest of living organisms, including humans, are **heterotrophs** and live off of the organic compounds produced by other organisms. Some kill to eat; others live on dead, decaying matter.

ESSENTIAL BACKGROUND

- **Electron transport chain (Chapter 7)**
- **Electromagnetic radiation (light)**

TOPIC 1: CHLOROPLASTS AND CHLOROPHYLL

KEY POINTS

✓ *What do chloroplasts contain? What do they do?*

✓ *What are thylakoid sacs?*

✓ *What is chlorophyll and where is it located?*

Chloroplasts are the green part of plants. Chloroplasts are several microns long (bacteria sized) and are the primary site for photosynthesis. Photosynthesizing cells, typically **mesophyll cells** in leaves, each contain about 30 to 40 chloroplasts. Chloroplasts, in turn, contain an inner **stroma** (fluid), stacks of **thylakoid sacs**, and two outer membranes. Each thylakoid sac is a membrane-bound disk. The membrane of the thylakoid sac is where chlorophyll is located, and is also the site where light is absorbed.

There are two stages to photosynthesis. The first, called the **light reactions (or photophosphorylation)**, converts light energy into the energy of chemical bonds in ATP and NADPH. **NADPH** is a close cousin to NADH, containing an additional phosphate group. During the light reactions, H_2O is split, and oxygen is formed. These reactions occur in the thylakoids. There are two different kinds of light reactions, **cyclic** and **noncyclic**, as we discuss in the next section.

The second stage of photosynthesis uses the free energy stored in the chemical bonds of ATP and NADPH to fix carbon atoms from carbon dioxide into organic compounds (carbohydrate). This second stage is called the **Calvin cycle** (or the **Calvin-Benson cycle**, or the **dark reactions**, since they can occur in the dark), and occurs in the interior stroma, of the chloroplast.

Photosynthesis is, roughly speaking, the reverse of cellular respiration. Electrons are moving uphill. In terms of energy content, free energy is increasing as the electrons in the chemical bonds of carbon dioxide and water become electrons in the chemical bonds of carbohydrates. This overall chemical reaction is endergonic, but photosynthesis couples it to sunlight as an energy source to power it.

Chlorophyll is the key to the initiation of the light reactions in the thylakoid membranes. Chlorophyll comes in two forms, a and b, which are minor chemical variants of one another. Chlorophyll is a modified lipid consisting of a multiple-ring structure containing mostly carbon and hydrogen atoms with a center magnesium atom. This structure is called a **porphyrin ring**, and is the light-absorbing end of the molecule. Chlorophyll also has a long hydrocarbon tail that helps to anchor the chlorophyll to the thylakoid membrane. Chlorophyll absorbs red and blue light. Green light is scattered or reflected, which gives plants their color. In the thylakoid membranes of chloroplasts, the chlorophyll molecules are clustered in **photosystems** with other pigment molecules, which broaden the light-absorbing spectrum. Photosystems are the light-harvesting units of the thylakoid membranes, and act as the collection point for light.

Topic Test 1: Chloroplasts and Chlorophyll

True/False

1. Chloroplasts directly link water and carbon dioxide together.
2. Light is absorbed during the Calvin cycle.
3. Photophosphorylation is another way of saying light reactions.

Multiple Choice

4. Which of the following is correct?
   a. Photoautotrophs are strict anaerobes.
   b. Photoautotrophs are heterotrophs.
   c. Heterotrophs live off of autotrophs, directly or indirectly.
   d. Chemoautotrophs are heterotrophs.
   e. Both a and d are correct.

5. During photosynthesis,
   a. oxygen is taken from the atmosphere.
   b. carbon dioxide is released to the atmosphere.
   c. carbohydrates are formed.
   d. plants get the energy they need, so they do not need mitochondria.
   e. Both a and c are correct.

Short Answer

6. What is the difference between chloroplasts and chlorophyll?

Topic Test 1: Answers

1. **False.** Water is split during the light reactions, whereas carbon dioxide is used in the Calvin cycle. There is no direct linking of water and carbon dioxide; however, the splitting of water generates hydrogen that can be used for carbohydrate production from carbon dioxide during the Calvin cycle.

2. **False.** Light is absorbed during the light reactions, as we explore in greater detail in the next section.

3. **True.** They often are used interchangeably, although we will see later that in reality, more than photophosphorylation occurs during the light reactions.

4. **c.** Photoautotrophs include plants, which are strict aerobes. Photoautotrophs, chemoautotrophs, and heteroautotrophs are each different from one another, as indicated in the introduction to the chapter.

5. **c.** Although plants get energy from sunlight, they need to make ATP for other cellular activities. The ATP that is made in chloroplasts is only used to make carbohydrates during the Calvin cycle.

6. Chloroplasts are cellular organelles, about the size of bacteria, that contain many chlorophyll molecules, which play a role in light absorption.

TOPIC 2: LIGHT REACTIONS

KEY POINTS

✓ *What are the two types of photosystems?*

✓ *Distinguish between cyclic and noncyclic photophosphorylation.*

✓ *What two compounds are generated by the light reactions?*

When the energy from light is absorbed by a center pair of chlorophyll molecules in a photosystem in the thylakoid membrane, an electron in the chlorophyll molecule is excited. The excited electron can be transferred to a nearby molecule, called the **primary electron acceptor**. The chlorophyll is oxidized, and the primary acceptor is reduced. There are two different kinds of photosystems in the thylakoid membrane: **photosystem I**, absorbing light best at a wavelength of 700 nanometers (the center of photosystem I is called P700), and **photosystem II**, absorbing light best at a wavelength of 680 nanometers (the center of photosystem II is called P680).

Two events can occur: cyclic photophosphorylation, which only involves photosystem I, and non-cyclic photophosphorylation, which involves photosystems I and II.

Cyclic photophosphorylation generates ATP. During this cyclic electron flow, energy from light absorbed by photosystem I releases an electron from a chlorophyll molecule, and that electron moves first to the primary acceptor and then down an electron transport chain. As the electron moves down the chain, it loses free energy and the free energy is used to pump protons (hydrogen ions) from the inner fluid of the chloroplast, the stroma, across the thylakoid membrane to the inside of the thylakoids. The protons pass back across the membrane through ATP synthase, generating ATP from ADP + P_i. At the end of the electron transport chain, the electron returns to the chlorophyll molecule, which is why this is called **cyclic electron flow**. Notice the similarities between this process and ATP production in mitochondria. Here, the inside of the thylakoids becomes acidic as protons are pumped.

Noncyclic photophosphorylation, or noncyclic electron flow, involves both photosystems, next to one another in the thylakoid membrane. We begin with light being absorbed by photosystem I, resulting in a transfer of excited electrons from chlorophyll to the primary acceptor. However, instead of passing all the way down the chain and generating ATP, a pair of these electrons is passed to $NADP^+$, forming, with the addition of a hydrogen ion, NADPH. This NADPH is similar to NADH in mitochondria, and contains about the same level of free energy in its bonded hydrogen.

Notice, however, that the electrons have been used to form NADPH, and so cannot return to their chlorophyll molecule (hence, the name noncyclic electron flow). Electrons are needed to fill the "holes" in the chlorophyll, and they come from photosystem II in the following manner: More light excites chlorophyll in photosystem II and electrons are excited enough to escape the chlorophyll and reach another primary electron acceptor. The electrons travel down an electron transport chain and fill the "holes" in the chlorophyll of photosystem I. As the electrons pass along the electron transport chain, protons are pumped across the membrane of the thylakoid, and ATP is synthesized through the resulting proton gradient. Notice that we have moved the electron "hole" from photosystem I to photosystem II. However, photosystem II has a special capability: It can generate the electron needed to fill the new hole by splitting water. The split water provides hydrogen ions, balancing those needed in the making of NADPH as well. Oxygen is given off as a by-product of this splitting of water. Notice that oxygen is produced in the light reactions, but no carbon dioxide has been used here. Carbon dioxide is used in the Calvin cycle, as is the energy from the NADPH and ATP that was formed during the light reactions. The combination of cyclic and noncyclic photophosphorylation ensures that ATP and NADPH can be made in the correct ratio for use in the Calvin cycle.

A summary of the light reactions is as follows:

$$\text{Light} + ADP + P_i + NADP^+ + H_2O \rightarrow ATP + NADPH + H^+ + O_2$$

This is not intended to be a balanced equation. Instead, it indicates the input and output for the light reactions. The ratio of ATP to NADPH produced by the light reactions can be altered by altering the ratio of cyclic to noncyclic electron flow.

Topic Test 2: Light Reactions

True/False

1. In cyclic electron flow, the inputs are light, ADP, P_i, and water.

2. In noncyclic electron flow, the only outputs are oxygen and ATP.

3. Some electrons from chlorophyll end up in NADPH molecules during the light reactions of photosynthesis.

Multiple Choice

4. Concerning ATP and NADPH synthesis in the light reactions of photosynthesis,
   a. ATP is made during both cyclic and noncyclic electron flow.
   b. NADPH is made during both cyclic and noncyclic electron flow.
   c. ATP is made only during cyclic electron flow.
   d. NADPH is made only during noncyclic electron flow.
   e. Both a and d are correct.

5. During the light reactions of photosynthesis,
   a. water is split and oxygen molecules are formed.
   b. the carbon in carbon dioxide is "fixed" into organic molecules.
   c. carbon-hydrogen bonds are formed.
   d. NADPH is split to form $NADP^+$.
   e. both b and d are correct.

6. Both during the light reactions in chloroplasts and during electron transport and oxidative phosphorylation in mitochondria,
   a. hydrogen ions (protons) get pumped across membranes.
   b. ATP is synthesized by ATP synthase.
   c. substrate-level phosphorylation occurs.
   d. water is split into oxygen.
   e. both a and b are correct.

Short Answer

7. What is the primary function of the light reactions in photosynthesis?

Topic Test 2: Answers

1. **False.** Water is not an input for cyclic electron flow.

2. **False.** NADPH + H^+ also are output.

3. **True.** This happens during noncyclic electron flow.

4. **e.** Do notice that ATP is made during both reactions, whereas NADPH is made, and water split, only during noncyclic photophosphorylation.

5. **a.** The other events occur during the Calvin cycle, as we discuss in the next section.

6. **e.** Both mitochondria and chloroplasts contain electron transport chains and make ATP using a proton gradient and ATP synthase.

7. The light reactions produce ATP and NADPH, which contain significant free energy in some of their chemical bonds. That free energy will be used in the Calvin cycle to produce organic molecules from inorganic molecules.

TOPIC 3: CALVIN (CALVIN-BENSON) CYCLE

KEY POINTS

✓ *How are carbohydrates (sugars) made during the Calvin cycle?*

✓ *What are the inputs and outputs of the Calvin cycle?*

Thanks to the light reactions, we have our energy source in the form of ATP and NADPH molecules. How do we make sugars from this energy? Just as there was a Krebs cycle for breaking down organic skeletons to carbon dioxide, there is a **Calvin**, or **Calvin-Benson cycle**, for building them up. The energy in NADPH allows us to form carbon-carbon and carbon-hydrogen bonds. We begin and end the cycle with a five-carbon sugar, **ribulose bisphosphate**. A carbon dioxide molecule is added to ribulose bisphosphate by the enzyme **rubisco**, and the new, six-carbon molecule splits into two identical three-carbon molecules, **3-phosphoglycerate.** In a series of enzyme-catalyzed steps, hydrolyzing ATP and oxidizing NADPH, **glyceraldehyde 3-phosphate** is formed. Additional enzymes use energy from more ATP to rearrange and combine glyceraldehyde 3-phosphate molecules to form ribulose bisphosphate once again.

The key to the output of carbohydrate from the Calvin cycle is the fact that for every three times through the cycle (adding a carbon atom from carbon dioxide each time), an extra glyceraldehyde 3-phosphate is formed. This extra glyceraldehyde 3-phosphate becomes an output of the cycle.

Two glyceraldehyde 3-phosphate molecules are the equivalent of one glucose molecule. To generate that equivalent of one glucose requires 18 ATP and 12 NADPH molecules, which are supplied by the light reactions. Six carbon dioxide molecules also were required. The hydrogens in the glucose were provided by the NADPH + H$^+$, and can be viewed as originally coming from the water that was split during the light reactions, when oxygen was formed. We have satisfied the overall reaction given in the introduction to this chapter. We have produced a molecule with 12 carbon-carbon and carbon-hydrogen bonds, and we have oxidized 12 NADPH molecules in doing so.

For information on **photorespiration**, **C$_4$** and **CAM plants**, see the web page.

Topic Test 3: Calvin Cycle

True/False

1. Rubisco is the enzyme that adds carbon dioxide to ribulose bisphosphate.

2. For every three carbon dioxide molecules taken into the Calvin cycle, one glyceraldehyde 3-phosphate molecule is generated.

3. The Calvin cycle hydrolyzes ATP and oxidizes NADPH.

Multiple Choice

4. Comparing the Krebs cycle and Calvin cycle,
   a. both generate ATP.
   b. both generate NADP (or NADPH).
   c. in the Krebs cycle, carbon dioxide is released, whereas in the Calvin cycle, it is incorporated into organic molecules.
   d. during the Krebs cycle, oxygen is converted to water, whereas during the Calvin cycle, water is converted to oxygen.
   e. All the above are correct.

5. Which of the following does NOT occur during the Calvin cycle?
   a. Carbon dioxide is used as a source of carbon atoms for making organic molecules.
   b. Water is split to form oxygen.
   c. Carbohydrate is made.
   d. ATP is hydrolyzed to ADP + P_i.
   e. NADPH is oxidized to $NADP^+$.

Short Answer

6. What do plants do with the output from photosynthesis?

Topic Test 3: Answers

1. **True.** We explore a different property of the enzyme rubisco in the next section.

2. **True.** This is the way that inorganic carbon is "fixed" into organic molecules.

3. **True.** This is the source of energy used to fix the carbon. The ATP and NADPH were generated during the light reactions.

4. **c.** The Calvin cycle does not generate ATP or NADPH; it uses them. Oxygen is not converted to water during the Krebs cycle but at the end of the electron transport chain.

5. **b.** Water is split to form oxygen during the light reactions, not during the Calvin cycle.

6. The carbohydrate output can be used by plants, together with minerals and amino groups from the soil, to generate the organic molecules that the plant needs. The most abundant of these usually is cellulose (Chapter 3), for cell walls. Some leaves generate sucrose (a disaccharide linking glucose and fructose) for export. Some plants store excess carbohydrates in the form of starch. In addition, the carbohydrate output can be used by plant mitochondria to generate ATP as an energy source for doing work in plant cells. The other major output of photosynthesis is oxygen, and some of it is released, contributing to the oxygen in our atmosphere. Some of it is used in plant mitochondria.

APPLICATION: PHOTOSYNTHESIS AND GLOBAL WARMING

Global warming is thought to be caused, in large measure, by increased carbon dioxide levels in the atmosphere (see Chapter 23). The additional carbon dioxide is being produced mostly by the burning of fossil fuels (oil, coal) by humans. Photosynthesis also plays a role in the level of carbon dioxide in the atmosphere. Photosynthesizing organisms fix carbon dioxide in organic molecules, lowering its level in the atmosphere. Even on an annual basis, the effect of such photosynthesis is measurable, since there is more carbon dioxide in the atmosphere of the Northern Hemisphere each winter, when there is less photosynthesis taking place.

It was hoped by some that higher carbon dioxide levels in the atmosphere would lead to an increase in photosynthesis that would help to prevent global warming by removing carbon dioxide from the atmosphere. Thus, the wood in trees contains fixed carbon in the form (mostly) of cellulose. If there is more wood in the world's forests, there will be less carbon dioxide in the atmosphere.

Unfortunately, we have been destroying forests at the same time that we have been burning fossil fuels. The destruction of forests reduces the amount of photosynthesis occurring, and contributes even more carbon dioxide to the atmosphere as the forests burn or decay. Thus, instead of giving worldwide photosynthesis a chance to reduce carbon dioxide levels, we are contributing even more carbon dioxide by destroying the forests.

There still is a chance that photosynthesis in the oceans might increase enough to help, but the increase in carbon dioxide levels in the atmosphere over the last 30 years indicates that any removal of carbon dioxide through photosynthesis has not been able to keep up with human production. If nothing is done, global warming will only get worse as human populations grow and as developing nations increase resource utilization.

Chapter Test

True/False

1. The chemical equation for photosynthesis is basically the opposite of the equation for cell respiration.

2. Plants are photoautotrophs; humans are chemoautotrophs.

3. Cyclic photophosphorylation is sometimes referred to as cyclic electron flow.

Multiple Choice

4. Which of the following is correct?
   a. Chlorophyll absorbs green light.
   b. The light reactions do not involve electron transport.
   c. During the light reactions, water is split, and oxygen is given off.
   d. NADPH is split during photophosphorylation.

5. During the Calvin cycle,
   a. there is a net loss of carbon-hydrogen bonds.
   b. light energy is converted into chemical energy.
   c. NADPH is formed.
   d. ATP is formed.
   e. none of the above is correct.

6. Photosystem I
   a. absorbs light best at the same wavelength as photosystem II.
   b. is not involved in making NADPH.
   c. can make NADPH through pumping of hydrogen ions across the thylakoid membrane.
   d. can gain electrons from photosystem II.
   e. can make ATP by substrate-level phosphorylation.

Chapter Test Answers

1. **T**   2. **F**   3. **T**   4. **c**   5. **e**   6. **d**

Check Your Performance:

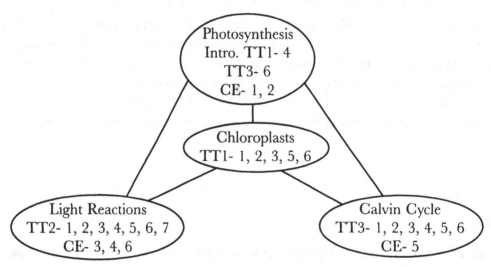

Key: TT = Topic Test; CE = Chapter Exam. Numbers indicate exam questions. Some questions are listed more than once if they refer to more than one topic.

Unit I Exam

Multiple Choice

1. An element, or kind of atom, can be determined by
   a. the number of protons in the nucleus of the atom.
   b. the number of neutrons in the nucleus of the atom.
   c. the sum of the number of protons plus neutrons in the atom.
   d. the density of the element.
   e. the color of the element.

2. Use the valences of (number of bonds formed by) the atoms to determine which of the following is most likely an independent molecule found in living organisms:

   a. $H_2N-\overset{\displaystyle H}{\underset{\displaystyle H}{\overset{|}{\underset{|}{C}}}}-OH$ b. $H=C=CH_2$ c. CH_3 d. $O=C-NH_2$

3. Which of the following is NOT one of the accepted, major generalizations in biology?
   a. All living things are made of cells.
   b. Living things evolve.
   c. Living things share such properties as growth, reproduction, organization, and use of energy.
   d. Living things exhibit vitalism, having special properties not governed by the laws of physics and chemistry.
   e. Genes specify proteins, which in turn specify metabolism and structure.

4. The bonding between hydrogen atoms and the oxygen atom in a single water molecule is best described as
   a. ionic bonding.
   b. nonpolar, covalent bonding.
   c. polar, covalent bonding.
   d. hydrogen bonding.
   e. none of the above.

5. The bonding between an oxygen atom on one molecule of water and hydrogen atoms on other, nearby water molecules in a water solution is best described as
   a. ionic bonding.
   b. nonpolar, covalent bonding.
   c. polar, covalent bonding.
   d. hydrogen bonding.
   e. none of the above.

6. Which of the following is most likely a carbohydrate?
   a. $C_{12}H_{26}O$
   b. $C_{12}H_{24}O_{12}$
   c. $C_{12}H_{24}O_2N_2$

d. $C_6H_6O_6$

e. $C_6H_6O_{12}$

7. How many asymmetric carbon atoms are present in the following molecule?

$$H-O-\underset{\underset{H}{|}}{\overset{\overset{H}{|}}{C}}-\underset{\underset{H}{|}}{\overset{\overset{H}{|}}{C}}=\overset{}{C}-\overset{\overset{O-H}{|}}{C}=O$$

a. One

b. Two

c. Three

d. Four

e. None

8. The tertiary structure of proteins is produced or stabilized by
   a. hydrogen bonds.
   b. ionic bonds.
   c. hydrophobic forces.
   d. disulfide bonds.
   e. all of the above.

9. Which of the following is the major lipid component of most biological membranes?
   a. Fats
   b. Phospholipids
   c. Cholesterol plus steroids made from cholesterol
   d. Carbohydrates

10. The total particle (ions plus molecules, etc.) concentration in a water solution outside of an animal cell is greater than that inside the cell. We can expect that
    a. the cell will shrink.
    b. the cell will stay the same size.
    c. the cell will swell.
    d. the cell will swell and burst.
    e. either c or d are correct, but we cannot tell which with the information given in the question.

11. Enzymes
    a. speed particular chemical reactions.
    b. increase the free energy difference between reactants and products.
    c. only aid endothermic reactions.
    d. do both a and b.
    e. do both b and c.

12. A membrane separates two solutions. Initially, on one side of the membrane is a 1.0 M sodium chloride solution, and on the other side is a 0.5 M potassium chloride solution. The membrane is permeable to chloride, but not to sodium or potassium. In the absence of water movement,
    a. the concentration of chloride will become equal on both sides.
    b. some chloride ions will move from one side to the other, but then a charge difference will prevent further net movement of chloride ions.

c. no net movement of chloride ions will occur.

d. there will be more chloride on the side with 0.5 M potassium than on the side with 1.0 M sodium, to offset the higher concentration of sodium.

13. Which of the following would necessarily violate the second law of thermodynamics?

a. A bacterium grows and divides into two, while consuming glucose from its environment.

b. The air pressure in Miami increases.

c. An air conditioner cools a room while using electricity and warming the air outside of the room.

d. Ice forms in a container of water, initially at a uniform temperature of 40°C, while the rest of the container gets warmer, and no other change occurs.

e. A balloon pops.

14. Which of the following functional groups is (are) found in proteins, but is (are) not commonly found in carbohydrates?

a. Sulfhydryl

b. Hydroxyl

c. Amino

d. Both a and c

e. All of the above

15. Fatty acid chains that have "kinks" in them are called unsaturated, and with more than one kink, they are called polyunsaturated. What causes these kinks?

a. Double or triple bonds between carbon atoms in the chains

b. Hydrogen bonds with nearby water molecules

c. Van der Waals forces among the chains

d. The presence of asymmetric carbon atoms in the middle of the chains

16. Dehydration (condensation) synthesis

a. can add a monomer unit to a growing chain.

b. produces a water molecule.

c. occurs during peptide bond formation.

d. occurs when glycogen is being formed.

e. does all of the above.

17. Substrate-level phosphorylation

a. describes how ATP is produced by the electron transport chain.

b. is the same as oxidative phosphorylation.

c. is the hydrolysis of ATP by enzymes.

d. is the moving of a phosphate group from a molecule to ADP, catalyzed by an enzyme.

e. includes all of the above.

18. What is the carrier of most of the electrons between the Krebs cycle and the electron transport chain?

a. ATP

b. ADP

c. Hydrogen

d. Oxygen

e. NADH

19. During cell respiration, oxygen is converted to water in which of the following?
    a. Glycolysis
    b. Krebs cycle
    c. Electron transport chain
    d. Both a and b
    e. Both a and c

20. ATP is generated from a proton (hydrogen ion) gradient during
    a. oxidative phosphorylation.
    b. the light reactions of photosynthesis.
    c. glycolysis.
    d. Calvin cycle.
    e. both a and b.

21. Carbon dioxide is either an input or an output of
    a. the light reactions of photosynthesis.
    b. the Krebs cycle.
    c. the Calvin cycle.
    d. the electron transport chain.
    e. both b and c.

22. ATP is generated by
    a. cyclic electron flow during the light reactions.
    b. noncyclic electron flow during the light reactions.
    c. glycolysis.
    d. Krebs cycle.
    e. all of the above.

23. In photosynthesis, oxygen is formed
    a. during the Calvin cycle.
    b. during photorespiration.
    c. during cyclic electron flow.
    d. during noncyclic electron flow.

24. (from WEB material) CAM and C_4 plants.
    a. are able to reduce photorespiration.
    b. incorporate carbon dioxide into organic acids.
    c. no longer run a Calvin cycle.
    d. do both a and b.
    e. do all of the above.

25. Concerning cell respiration, which of the following would eventually block the synthesis of ATP by oxidative phosphorylation?
    a. Blocking the flow of electrons along the electron transport chain
    b. Blocking the movement of protons through ATP synthase
    c. Making the inner mitochondrial membrane very permeable to hydrogen ions
    d. All of the above

Short Answer

26. Complete the blanks in the following table, which describes a water solution:

$[H^+]$	$[OH^-]$	pH	Acidic, Basic, or Neutral?
	10^{-9} M		

27. a. What monomer unit is used to synthesize the carbohydrate polymer glycogen?
    b. What monomer unit is used to synthesize the plant cell-wall polymer, cellulose?
    c. What is the primary chemical difference between glycogen and cellulose?

28. List the net products of glycolysis per molecule of glucose.

29. During fermentation, some _____ is produced, which gives the cell energy in a usable form. In addition, depending on the type of fermentation, either _____ or _____ is produced. During formation of these products, _____ recycles to _____, which allows fermentation to continue.

30. a. Which cytoskeletal component is most important in chromosome movements during mitosis?
    b. Which cytoskeletal component is also used in muscle cell contraction?

Unit I Exam Answers

1. **a** 2. **a** 3. **d** 4. **c** 5. **d** 6. **b** 7. **e** 8. **e** 9. **b** 10. **a** 11. **a** 12. **b**

13. **d** 14. **d** 15. **a** 16. **e** 17. **d** 18. **e** 19. **c** 20. **e** 21. **e** 22. **e** 23. **d**

24. **d** 25. **d**

26. $[H^+] = 10^{-5}$M; pH = 5; acidic

27. **a**. Glucose **b**. Glucose **c**. Glycogen uses alpha-glucose; cellulose uses beta-glucose. These two forms of glucose are enantiomers

28. 2 pyruvate + 2 ATP + 2 NADH + 2H$^+$

29. ATP; ethanol (ethyl alcohol); lactate (lactic acid); NADH; NAD$^+$

30. **a**. Microtubules **b**. Microfilaments (actin filaments)

UNIT II:
GENES,
INFORMATION,
AND HEREDITY

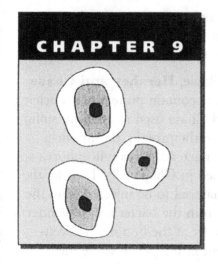

CHAPTER 9

DNA Structure, Function, and Replication

Scientists first defined **genes** as units of inheritance. They then realized that there was a correlation between the rules governing chromosome distribution during cell division and Mendel's rules governing inheritance of traits. Thus, it was presumed that chromosomes contained the genetic material, but was it the proteins or the DNA of the chromosomes? As we will see, two important experiments identified DNA as the genetic material. Shortly afterward, the structure of DNA was determined, and that structure suggested how information was stored in DNA and how the DNA might be copied. We have since confirmed that DNA contains the genetic blueprints for the construction of organisms. The ability to make copies of the DNA allows genetic information to be passed from parents to offspring, as well as for the same information to be present in the many cells of multicellular organisms.

ESSENTIAL BACKGROUND

- **Nucleotides and bases (Chapter 3)**
- **Deoxyribose sugar (Chapter 3)**
- **Hydrogen bonds (Chapter 2)**
- **Isotopes (Chapter 2)**
- **Density Gradients**

TOPIC 1: EVIDENCE THAT DNA IS THE GENETIC MATERIAL

KEY POINTS

✓ *What experiments first demonstrated that DNA was the genetic material?*

✓ *Who performed those experiments, and when?*

✓ *How was the possibility that protein was the genetic material ruled out?*

In 1928 **Griffith and coworkers** demonstrated the occurrence of bacterial **transformation** (an addition of genetic material to bacteria—see Chapter 14). Mice died after they were injected with living S-strain pneumococcus (*Streptococcus pneumoniae*; these bacteria cause pneumonia). Mice treated with heat-killed S-strain, or with R-strain, a different genetic strain, lived. However, if the mice were treated with a mixture of the heat-killed S-strain plus the R-strain bacteria, the mice died. The dead mice contained living S-strain bacteria. It was concluded that some of the R-strain bacteria were transformed to the killer S-strain by genetic material from the heat-killed

S-strain bacteria. In 1944 **Avery and coworkers** purified the transforming material that converted R-strain to S-strain and found that the material purified just like DNA did, not like protein did. This was strong evidence in favor of DNA as the genetic material.

Shortly after World War II, when radioactive isotopes became available, **Hershey and Chase** identified the part of a virus that contained genetic material. Viruses contain protein and nucleic acid (DNA or RNA, depending on the kind of virus). Hershey and Chase used radioactive sulfur to selectively tag the proteins (sulfur is found in some amino acids in the protein, but not in DNA), and radioactive phosphorus to selectively tag the DNA in a bacterial virus (phosphorus is very abundant in DNA and not so abundant in proteins, as discussed in Chapter 3). They let the virus bind to the bacteria, allowed a few minutes for the genetic material to be injected from the virus into the bacteria, and then broke the coat of the virus away from the bacteria in a blender. After centrifuging the infected bacteria and collecting them in a pellet at the bottom of a centrifuge tube, they found that the radioactive phosphorus was present in the pellet, while the radioactive sulfur was still suspended, in the empty coats of the virus, which had remained in the supernatant because they were smaller than the bacteria. Thus, it was the DNA that was present in the infected bacteria, while the protein was in the virus coat, which had broken away. This gave further strong support to the hypothesis that DNA was the genetic material.

By 1950, because of the Griffith-Avery and Hershey-Chase experiments, most scientists were convinced that DNA is the genetic material. Since that time many experiments have confirmed this hypothesis, strengthening it to a well-supported theory at the very foundation of modern molecular biology.

Topic Test 1: Evidence That DNA Is the Genetic Material

True/False

1. Chromosomes contain protein and DNA.

2. Griffith and Chase together performed an experiment that demonstrated that DNA is the genetic material.

3. Hershey and Chase together performed an experiment that demonstrated that protein is the genetic material.

Multiple Choice

4. In the Hershey-Chase ("blender") experiment,
   a. the radioactive phosphorus selectively labeled the protein.
   b. the radioactive sulfur selectively labeled the DNA.
   c. protein was shown to be the genetic material.
   d. the genetic material of the virus was removed from the bacteria by putting the infected bacteria in a blender.
   e. None of the above is correct.

5. In the Griffith-Avery experiments on the nature of the genetic material, which of the following is NOT correct?
   a. A transforming factor from a heat-killed strain of bacteria was determined to purify like DNA.

b. Live R-strain bacteria, mixed with heat-killed S-strain, were able to kill mice, while remaining R-strain.

c. Some of the live R-strain, when mixed with heat-killed S-strain bacteria, became S-strain bacteria.

d. The heat-killed S-strain must have contained genetic material that got transferred, somehow, to the R-strain bacteria.

Short Answer

6. Why was sulfur used as a radioactive label by Hershey and Chase?

Topic Test 1: Answers

1. **True.** Chromosomes also contain some other molecules, such as RNA, but proteins and DNA make up the bulk of the mass of chromosomes. Once it was realized that chromosomes contained the genetic material, the main task of identifying the genetic material involved determining whether it was protein or DNA.

2. **False.** Griffith and Avery performed a set of experiments that differed from the Hershey and Chase experiments.

3. **False.** DNA, not protein, was identified as the genetic material.

4. **e.** None of the other answers are correct. Phosphorus labeled the DNA; sulfur labeled the protein; DNA was shown to be the genetic material; and the protein coat was removed by the blender—the virus's genetic material stayed inside of the bacteria, since the bacteria remained infected.

5. Answer b is not correct. Some of the R-strain bacteria were transformed to S-strain, and these killed the mice.

6. Sulfur was used to label protein because two amino acids found in protein contain sulfur, whereas sulfur is not present in DNA (see Chapter 2 for a description of the elements found in DNA and protein).

TOPIC 2: DNA STRUCTURE

KEY POINTS

✓ *What data helped Watson and Crick determine the structure of DNA?*

✓ *What is the structure of DNA?*

✓ *What holds the two strands of the double helix in place?*

✓ *What does the structure of DNA suggest about its function?*

Watson and Crick determined the structure of DNA in 1953. They were aided by two sets of data from other scientists. As is often the case in science, new knowledge builds on the foundation of the work of others. **Chargaff and coworkers** had shown that the four bases, which distinguish the four kinds of nucleotides found in DNA, were present in particular ratios in the DNA from all of the different kinds of organisms that they examined. The percent of adenine (A) was always equal, within error, to the percent of **thymine (T)**, and the percent of **guanine (G)** was always equal, within error, to the percent of **cytosine (C)**. These are called **Chargaff's rules**: %A = %T and %G = %C.

The other data used by Watson and Crick consisted of x-ray diffraction patterns for DNA from **Rosalind Franklin** who was working in **Maurice Wilkins's** laboratory. The patterns indicated that DNA was a helix with certain repeat spacings. Rosalind Franklin died of cancer before the Nobel Prize was awarded to Watson, Crick, and Wilkins.

Watson and Crick proposed that DNA consisted of two strands, running in opposite directions from one another (**antiparallel**), with the bases of the nucleotide pairs linking the two strands by hydrogen bonds. Each strand consisted of a chain of nucleotides, linked directly to one another. You will recall that each nucleotide consists of a phosphate group linked to a deoxyribose sugar, which in turn is linked to a nitrogenous base (Chapter 2). Along each strand of the DNA molecule, the phosphate group in each nucleotide is linked to two sugars, one its own and the other its neighbor, forming a repeating backbone of sugar-phosphate-sugar-phosphate-sugar-phosphate. The nitrogenous base in each nucleotide projects from the backbone toward a base on the other strand. Two strands make up one DNA **double helix**. The DNA looks rather like a twisted stepladder, with pairs of bases, one from each strand, serving as the rungs on the ladder. Watson and Crick also inferred from the x-ray diffraction data that the width of the helix was 2 nanometers (10^{-9} meters), the distance between the base-pairs (steps on the ladder) was 0.34 nanometer, and the repeat distance for one turn of the helix was 3.4 nanometers, or 10 base-pairs. The two strands are held in register because the sequence of bases on the two strands are **complementary**: A on one strand always bonds with T on the other, and G with C. That pairing of the bases gives rise to Chargaff's rules. The pairs of bases that link across from one strand to the other are held in register with hydrogen bonds (Chapter 2) that form between each pair. Thus, the same kind of weak bonds that hold water molecules together also hold the two strands of DNA together.

The Watson and Crick model has now been supported by a wide range of evidence. The idea that the sequence of bases along a strand (which might be, for example, AGGTTACAC-CCAGTTAC, etc.) contains genetic information now has very full support, as described in the next chapter. A single DNA molecule can be quite long, containing millions of nucleotides linked in a specific sequence along each strand. Because of the base pairing in a DNA molecule, the sequence of nucleotide bases along one strand can be used to predict the sequence along the opposite, so-called complementary, strand by using Chargaff's rules. For example, the sequence ATCGGT on one strand links to the complementary sequence TAGCCA on the other strand. Human cells contain 46 long DNA molecules, each one running the length of a chromosome. In total they contain 50,000 to 100,000 different genes. If stretched out and put end-to-end, the DNA from a single human cell would be more than 1 meter long!

Topic Test 2: DNA Structure

True/False

1. Watson and Crick used Chargaff's x-ray diffraction patterns to determine the structure of DNA.

2. Each DNA molecule contains two sugar-phosphate backbones in its center, with nucleotide bases pointing outside.

3. The two strands of linked nucleotides in a single DNA molecule run in opposite directions, and so are referred to as being antiparallel.

Multiple Choice

4. Which of the following does NOT follow from Chargaff's rules?
   a. %A = %T
   b. %G = %C
   c. %A = %C
   d. (%A + %G) = (%T + %C)

5. If the sequence of nucleotide bases on one strand of a DNA double helix is GATTACA, the sequence on the opposite strand is
   a. GATTACA
   b. CTAATGA
   c. CUAAUGA
   d. AGCCGTG

Short Answer

6. We will see in the next chapter that about 1,500 nucleotides are required for the information needed by a typical gene. Given the distance between nucleotide pairs, and given the total length of DNA in a single human cell as at least 1 meter, calculate the possible number of genes in human DNA. Compare that to the actual number of 50,000 to 100,000.

Topic Test 2: Answers

1. **False.** The x-ray diffraction patterns were those of Rosalind Franklin.

2. **False.** The sugar-phosphate backbones are on the outside of the double helix, with the bases paired on the inside.

3. **True.** Each of the two strands can be referred to by the free phosphate-linkage points at the ends of each DNA strand. These are at carbons 5′ and 3′ on the ribose sugars. If one of the strands runs from 3′ to 5′, the other strand runs from 5′ to 3′, in the opposite direction.

4. **c.** DNA from different sources can have different ratios of A to C. All of the rest of the answers follow from Chargaff's rules.

5. **b.** Answer a repeats the same sequence, and would result in a DNA that would not follow Chargaff's rules. Answer c is actually the sequence that an RNA molecule would have if copied off of the DNA strand (U replaces T), as we will see in the next chapter. Answer d has A pairing with G and T with C, which would result in a DNA molecule not following Chargaff's rules.

6. The distance between nucleotide pairs is 0.34 nanometer. Therefore, in a meter-long set of DNA molecules, the number of such pairs is about 2.9×10^9, or 2.9 billion. In contrast, 100,000 genes times 1,500 nucleotides per gene gives only 150 million

nucleotides. We actually need to account for the two copies of most genes in each cell, which would double the number of nucleotide pairs to 300 million. Even with this adjustment, it appears that the amount of DNA in the human genome is five to 10 times more than is needed to specify the genes for humans. The rest of the DNA has a number of uses, as we will see in the next chapter.

TOPIC 3: DNA REPLICATION

KEY POINTS

✓ *What does the structure of DNA suggest about how it gets copied?*

✓ *When DNA gets copied, what happens to the existing strands of DNA?*

✓ *What does semiconservative replication mean?*

✓ *What replicates DNA and what are the details of the process?*

✓ *What are Okazaki fragments and why are they produced during the copying of DNA?*

Watson and Crick realized that the double-stranded nature of DNA provided an easy way for the information to be copied: as the double strands unwind, each strand could be used as a guide, or template, to make a copy of the other, complementary, strand. If the process of unwinding and copying each strand were to continue, one would end up with two identical DNA molecules, each a double helix. Each of the DNA molecules would contain one old and one new strand. This is called **semiconservative replication**. Thus, Watson and Crick's proposed structure suggested an answer to the question of how the DNA gets replicated.

But how could it be demonstrated that DNA is copied in this way? **Meselson** and **Stahl** made use of isotopes to follow the process of the copying of DNA in bacteria. By shifting from heavy to light isotopes in the nutrients that the bacteria were growing on, Meselson and Stahl were able to change the density of the DNA being made by the bacteria. They compared the density of the DNA one and two generations after the shift in isotopes in the bacteria, by applying DNA from the bacteria to a density gradient in a centrifuge. After one generation, all of the DNA was of intermediate density—exactly as expected if each of the new DNA molecules contained one light and one heavy strand (one old and one new strand), as predicted by semiconservative replication. If the two strands in the old DNA molecule had stayed together, and the two new ones had been linked together in a totally new molecule, one would have conservative replication, and there would have been two densities for the DNA, one light and the other heavy, with no DNA of intermediate density. Thus, the results of the Meselson-Stahl experiment showed that replication of DNA is semiconservative, and not conservative.

After two generations, Meselson and Stahl found that one-half of the DNA was light in density, and the other half was of intermediate density, again as can be predicted from semiconservative replication.

Since this experiment was completed, our knowledge of the details of DNA replication has grown. DNA does not copy itself. Instead, proteins catalyze the various steps in the process, as shown in **Figure 9.1**. One class of proteins, called **helicases**, unwinds the double helix ahead of the main **DNA polymerase** molecule, which catalyzes the addition of nucleotides to the new, growing strand, using one of the existing strands as a template. This DNA polymerase can only copy in one direction along the DNA strand, making a strand that starts from its 5′ end and

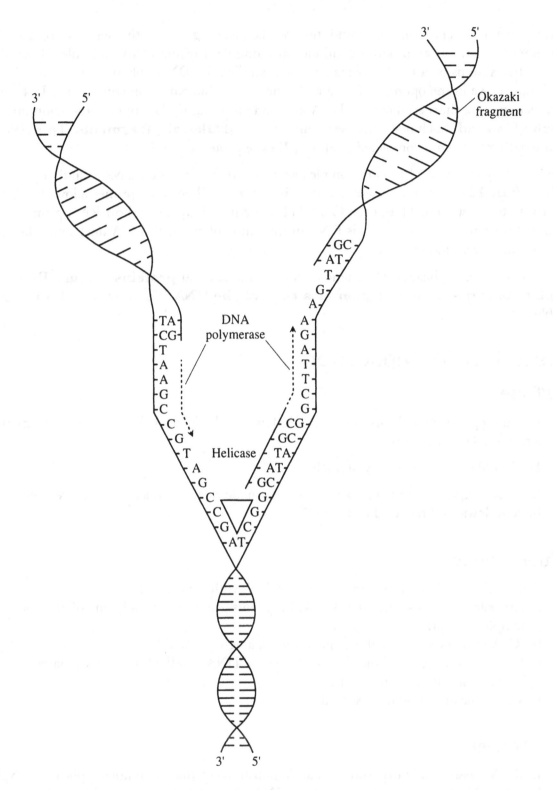

Figure 9.1 DNA replication. Each DNA polymerase molecule "reads" the template strand as it catalyzes the construction of a new strand. The helicase unwinds the DNA. The DNA polymerases run in opposite directions on the two strands.

builds toward the 3′ end. On one strand, the so-called leading strand, this synthesis of a new, complementary strand can progress continuously along the existing DNA molecule. However, since the two strands of a DNA molecule are antiparallel, the DNA polymerase on the other strand must move in the opposite direction. As the helicase advances, unwinding the DNA in one direction, new, short segments of DNA are made running in the opposite direction on this **"lagging" strand**. Such new, short segments are called **Okazaki fragments**. These get linked together to form a completed strand by **ligase** proteins.

Things actually are even a bit more complex since at the start of each DNA strand, a short **"primer" of RNA**, several nucleotides long, is first made. The RNA primer allows the DNA polymerase to get started. Thus, each Okazaki fragment initially has a short RNA primer attached. Other proteins strip out this RNA primer and "fill in" with DNA nucleotides before the ligase links everything together.

The mechanism of replication of genetic material indicates that **proteins can use DNA as a template to catalyze construction of a copy of the DNA**. This is what makes heredity possible.

Topic Test 3: DNA Replication

True/False

1. During replication of DNA, a new DNA double helix is formed, with two newly made strands linked to each other.

2. DNA is able to make a copy of itself.

3. Okazaki fragments are temporary molecules made during replication and are quickly broken down and replaced by new DNA.

Multiple Choice

4. Which of the following statements about DNA synthesis is NOT correct?
   a. At a site of DNA synthesis, Okazaki fragments are made on only one of the two template strands.
   b. RNA primers are needed to initiate the making of new DNA.
   c. DNA ligase is used to link the two strands of DNA together, across base-pairs.
   d. The two strands are antiparallel.
   e. The replication is semiconservative.

Short Answer

5. In the Meselson-Stahl experiment that demonstrated semiconservative replication, explain how the observed density distribution in DNA arises after two generations.

6. What role does the existing strand of DNA play during the synthesis of a new strand?

Topic Test 3: Answers

1. **False.** After replication, each DNA molecule contains one new strand and one old strand. The two newly made strands are not part of the same molecule.

2. **False.** DNA cannot copy itself. Instead, proteins catalyze the construction of a copy of the DNA.

3. **False.** Okazaki fragments are not broken down but get linked together by ligase to form a new strand of DNA.

4. **c.** Ligase does not link the two strands together. The two strands are only held together by hydrogen bonds between the nucleotide bases. The ligase links parts of one strand together.

5. At the end of one round of replication (DNA synthesis), each DNA molecule contains one old (heavy isotope) chain and one new (light isotope) chain. Thus, all DNA is of intermediate density. After a second round of replication, the strands will separate and each will become linked to a new (light isotope) strand. Thus, one all-light DNA will be formed and one heavy-plus-light DNA will be formed. So, half of the DNA will have a light density and the other half will have an intermediate density.

6. The existing strand is used as a template, or guide, that specifies for the DNA polymerase what sequence of nucleotides to add to the newly forming strand. Thus, the sequence of nucleotides on the existing strand specifies, through Chargaff's rules, the sequence of nucleotides in the new, complementary strand.

APPLICATION: THE HUMAN GENOME PROJECT

An attempt is underway to determine the entire nucleotide sequence of the DNA in all 46 human chromosomes. Each chromosome contains a single, long molecule of DNA. The project is challenging because of the total number of nucleotides involved—almost 3 billion. A number of laboratories in several different countries are involved in the task, which should be completed sometime early in the next century. It is hoped that knowing the sequence of nucleotides in the human genome will be helpful in detecting, and perhaps correcting, human genetic defects, as well as in determining the nature of various genes in humans.

The entire genome sequences of several bacteria and over 100 viruses have already been determined. A complete sequence also exists for yeast, a single-celled organism containing 12 million nucleotide base-pairs in its DNA. Scientists just recently completed a determination of the entire sequence for a nematode, a multicellular animal whose genome contains about 100 million nucleotide base-pairs.

Chapter Test

True/False

1. Hershey and Chase concluded that DNA was the genetic material because radioactive phosphorus remained with the bacteria after the virus coat was broken away in a blender.

2. The copying of a DNA molecule in a living cell requires a number of different kinds of proteins.

3. The ratio of guanine to adenine in DNA from different species can differ.

Multiple Choice

4. Concerning the structure of DNA, which of the following is false?
   a. DNA is a double helix with a 0.34-nanometer repeat distance between base-pairs along the helix.
   b. We have only known the structure of DNA since 1953.
   c. In DNA, opposite each guanine base is a cytosine base.
   d. Along the backbone of each strand of a DNA molecule (excluding the very ends of the molecule), each phosphate group is directly, covalently bonded to only one deoxyribose sugar.

5. Concerning DNA synthesis, which of the following is true?
   a. Large carbohydrates play an important role in DNA synthesis.
   b. The existing DNA double helix does not unwind while a copy of the DNA is made.
   c. In DNA synthesis, first an entire copy of the DNA is made with RNA, and then the RNA is converted to DNA.
   d. DNA can make a copy of itself with the proper triphosphate building blocks, with no other macromolecules being involved.
   e. None of the above is correct.

6. In a DNA molecule, which of the following pairs of chemical structures is held together by hydrogen bonds?
   a. Ribose-uracil
   b. Sugar-phosphate
   c. Adenine-thymine
   d. Phosphate-cytosine
   e. Deoxyribose-guanine

Short Answer

7. One strand of a DNA has the sequence AAGCCTTTG. What sequence is on the other strand?

*8. Fifteen percent of the nucleotide bases in a DNA molecule are cytosine. What can be said about the amount of guanine, thymine, and adenine in this DNA?

Chapter Test Answers

1. **T** 2. **T** 3. **T** 4. **d** 5. **e** 6. **c**

7. The other strand will have the sequence TTCGGAAAC, which can be determined by following Chargaff's rules.

8. If 15% is cytosine, then there must be 15% guanine, by Chargaff's rules. This totals 30%. The rest of the DNA, or 70%, must be equal amounts of adenine and thymine, so these are 35% each.

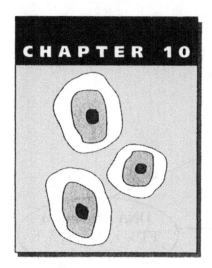

RNA and Protein: Transcription, Translation, and the Genetic Code

In the last chapter we learned about the structure of DNA and how DNA is copied. Now we will see how the genetic information in DNA is used to make proteins. This gets us to the heart of molecular biology. The so-called **central "dogma" of molecular biology** relates to the flow of genetic information: **DNA → RNA → protein**. The first arrow represents the process of transcription, the flow of information from DNA to RNA. The second arrow represents translation, the flow of information from RNA to protein. When we consider certain kinds of viruses in Chapter 14, we will be making an addition to this model of the way that information flows.

Much of the information in DNA is in the form of a code. In the language of DNA, words are called **codons**, and the alphabet consists of only four letters (four nucleotides). Codons are three letters long, and most codons specify a particular one of the 20 amino acids used by cells to build proteins. This chapter describes how this genetic information is used to make proteins. The story begins with the making of RNA molecules, which play a variety of roles in protein synthesis.

ESSENTIAL BACKGROUND

- Nucleotides and nucleic acids (Chapter 3)
- Differences between DNA and RNA (Chapter 3)
- Chargaff's rules (Chapter 9)
- Peptide bonds (Chapter 3)

TOPIC 1: TRANSCRIPTION

KEY POINTS

✓ *How is RNA synthesized in cells?*

✓ *What is the role of DNA in RNA synthesis?*

✓ *What is the sequence of steps involved in transcription?*

RNA plays a variety of roles in protein synthesis. There are three important kinds of RNA, and all three are made by a similar mechanism. Information in DNA is used to make all three

Check Your Performance:

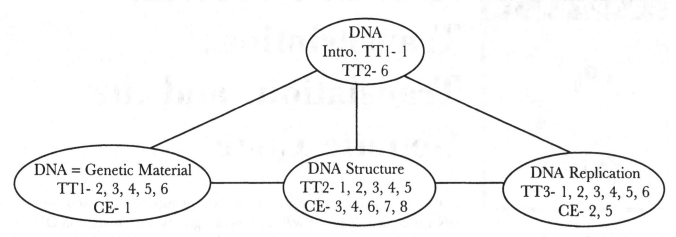

Key: TT = Topic Test; CE = Chapter Exam. Numbers indicate exam questions. Some questions are listed more than once if they refer to more than one topic.

kinds—the sequence of bases along parts of one strand of a DNA molecule is used to specify the sequence of bases in the RNA. This is similar to the way that DNA replication uses the sequence of nucleotide bases along one strand to specify the making of a new DNA strand. DNA serves as the **template** from which the RNA is made. However, RNA synthesis is a simpler process than DNA synthesis because only one strand of the DNA needs to serve as a template for the making of RNA. The process of making an RNA molecule on a template of DNA is called **transcription**.

RNA polymerase is the enzyme that transcribes RNA from a DNA template. In eukaryotic cells there are different kinds of RNA polymerase for the different kinds of RNA, but the general process of transcription appears to be the same for all types of RNA. The RNA polymerase links nucleotides together to make an RNA molecule, and uses the sequence of nucleotides on one strand of the DNA to specify a complementary sequence on the RNA. The usual relationships hold between the sequence of bases in the DNA template and the resulting RNA base sequence, except that in RNA, **uracil (U)** plays the role of thymine (T). Thus, the RNA polymerase links a U to the RNA chain whenever A is the next base on the template strand of the DNA, links A whenever T is next on the DNA, links G when C is next, and links C when G is next. Thus, the sequence ATGGCC in the coding, or template, strand of the DNA would be transcribed to UACCGG in the RNA. ATP, UTP, CTP, and GTP are the precursors used by the RNA polymerase, with diphosphate groups being hydrolyzed from the triphosphates, which provides free energy for the linking of the remaining nucleotides to the growing RNA chain.

The steps in transcription are as follows:

1. Binding. RNA polymerase binds to the DNA at particular sequences along the DNA. The binding regions are called **promoters**. Special transcription factors, generally proteins, can control the binding of RNA polymerase and the initiation of RNA synthesis. Some of these transcription factors bind directly to the DNA and can promote or inhibit the binding of RNA polymerase or the initiation of RNA synthesis. Other such factors bind directly to the RNA polymerase.

2. Initiation. The DNA double helix unwinds in a small region, and initiation of synthesis of RNA begins at a specialized initiation site near the promoter region.

3. Elongation. The RNA polymerase moves along one DNA strand making a complementary RNA strand, which separates from the DNA as the DNA double helix rewinds when RNA polymerase moves along. The RNA chain grows from the 5′ end toward the 3′ end (Chapter 9), just as DNA strands do.

4. Termination. A particular signal sequence along the DNA indicates that the end of the RNA molecule has been reached. RNA polymerase ceases synthesis and separates from the DNA, as does the completed RNA molecule.

5. Processing. After synthesis there is further modification to some RNA molecules, as we discuss later in this chapter.

Transcription produces a single-stranded RNA, whose sequence of bases has been specified by the part of the DNA molecule from which it was transcribed. A single-stranded RNA molecule often can fold back on itself in places, forming segments of double helix whenever the folding creates (usually short) chains of complementary base-pairs. This gives the otherwise randomly coiled RNA molecules some more stable, secondary structure.

Topic Test 1: Transcription

True/False

1. The binding of RNA polymerase and the initiation of transcription can involve special transcription factors that regulate the rate at which RNA is made.

2. Codons are three nucleotides long.

3. All RNA molecules are terminated after exactly 1,000 nucleotides.

4. RNA is double-stranded, as is DNA.

5. In the central dogma, DNA → RNA → protein, the arrows represent the flow of information.

Multiple Choice

6. A DNA molecule has the sequence ATTGAACCC along one strand that is being transcribed by an RNA polymerase enzyme. The resulting RNA molecule will have the sequence
   a. ATTGAACCC.
   b. TAACTTGGG.
   c. UAACUUGGG.
   d. GCCAGGTTT.
   e. none of the above.

Short Answer

7. Why are the binding and initiation phases efficient ones for the regulation of transcription by transcription factors?

Topic Test 1: Answers

1. **True.** This is a major point for the regulation of gene expression.

2. **True.** The genetic code contains "words" that are three letters long.

3. **False.** Termination of transcription is controlled by special base sequences in the DNA.

4. **False.** RNA usually is single-stranded. Double-stranded RNA is relatively rare, being found in some viruses and virus-infected cells.

5. **True.**

6. **c.** In RNA, the base uracil (U) substitutes for thymine (T) in DNA. Thus, the matches that are made during RNA synthesis from a DNA template, going from DNA to RNA, are A to U; T to A; G to C; and C to G.

7. These phases are more efficient than regulation that occurs later because the cell does not expend energy (nucleotide triphosphate hydrolysis) to make unneeded RNA molecules.

TOPIC 2: KINDS OF RNA AND RIBOSOME CONSTRUCTION

KEY POINTS

✓ *What are the three principal types of RNA?*

✓ *What roles do each of the three kinds of RNA play in protein synthesis?*

✓ *What are ribosomes composed of, and how are they constructed?*

✓ *What are introns and exons?*

There are three main kinds of RNA:

1. Ribosomal RNA, or **rRNA**. There are several different varieties of rRNA molecules in a cell; most are several thousand nucleotides long and all become parts of ribosomes, which are key players in protein synthesis.

2. Transfer RNA, or **tRNA**. These short (75–85 nucleotides long) RNA molecules have single amino acids linked to them, and play a role in then linking the amino acids to the growing polypeptide chain during protein synthesis on the ribosome.

3. Messenger RNA, or **mRNA**. There are a variety of mRNA molecules in any given cell, and each carries the information specifying the sequence of amino acids for a polypeptide. This is the molecule that carries the genetic information from DNA to protein. It is the RNA in the famous information flow formula from molecular biology: DNA → RNA → protein.

All of these RNA molecules are made from DNA as a template, as described earlier. There are other kinds of RNA, and these will be mentioned as the properties of the "big 3" are described below.

Ribosomes contain two subunits, large and small. Each subunit contains one or more rRNA molecules plus a variety of proteins. In the bacteria, *Escherichia coli*, a typical prokaryotic organism, each ribosome contains a total of three different rRNA molecules and 55 proteins. Eukaryotic ribosomes are somewhat more complex, containing four rRNA molecules and about 70 proteins. As discussed in Chapter 5, ribosomes are put together in the nucleolus, a specialized region of the nucleus. The nucleolus contains the DNA that has the genetic information for making rRNA molecules. The proteins that become part of the ribosome are made in the cytosol, outside of the nucleus, and then are transported into the nucleus. Small and large ribosomal subunits are shipped from the nucleus to the cytoplasm, where they are used for protein synthesis, as we will see.

The ribosome is a rather complicated structure, composed of about 60 to 80 macromolecules. It can be put together by self-assembly. The structure of the parts contains all of the information necessary to form the completed ribosome structure—a self-contained blueprint. Molecular-level forces serve as glue and fasteners.

Transfer RNA molecules are the carriers for amino acids. There are different kinds of tRNA molecules, and each one accepts (gets charged with) only one of the 20 amino acids used in protein synthesis. Enzymes called **aminoacyl-tRNA synthetases** do the linking of particular amino acids to particular tRNA molecules. The enzymes hydrolyze ATP as they make the linkage. The tRNA-to-amino-acid bond is an energetic one, providing energy to form the peptide bond between amino acids in the protein. Aminoacyl-tRNA synthetases are a key to the accuracy of protein synthesis. If the wrong amino acid were to be added to a tRNA, there is no correction mechanism—it will be inserted into the wrong place in the growing polypeptide chain. Fortunately, the aminoacyl-tRNA synthetases are very accurate.

Transfer RNA molecules also contain anticodons, which are triplets of nucleotides that will "recognize" codon triplets on the mRNA during protein synthesis, as we describe later.

Messenger RNAs are information carriers. The sequence of nucleotides in the mRNA specifies the sequence of bases in protein. The genetic code gives us the translation between base triplet codons in the mRNA and amino acids in the protein. In eukaryotic organisms, but not in prokaryotic organisms, the transcribed pre-mRNA (or hnRNA—heteronuclear RNA) molecules that become mRNA are made as larger molecules and then cut down to size. Present in the freshly transcribed pre-mRNA molecule are segments called **introns** that are spliced out during processing in the nucleus to form mRNA. The segments that get linked together to form the mRNA are called **exons**. It may help you to remember the difference between introns and exons if you remember the introns stay *in* the nucleus, and exons *exit* the nucleus.

The splicing out of introns is done by special particles that contain a fourth kind of RNA, called **small nuclear RNA**, or **SnRNA**, that has catalytic activity. Catalytically active RNA molecules are called **ribozymes**, a counterpart to protein enzymes. There are a variety of SnRNAs that splice RNA at particular nucleotide sequences. These splicings need to be very accurate so as not to disrupt the information in the mRNA.

Messenger RNA molecules are further processed in the nucleus with a chemical modification of a GTP on its 5′ end, called a **5′ cap**, and the addition of a set of adenine nucleotides to the 3′ end. The latter set of nucleotides is called a **poly-A tail**. It seems to play a role in setting the lifetime of the mRNA—the longer the tail, the longer the half-life of the mRNA. Each mRNA leaves the nucleus and is linked to ribosomes in the cytosol to specify the amino acid sequence in a protein.

Topic Test 2: Kinds of RNA and Ribosome Construction

True/False

1. Messenger RNA molecules are linked introns.

2. Ribosomes contain proteins plus RNA.

3. SnRNAs are a fourth kind of RNA.

Multiple Choice

4. Which of the following is a shorter RNA molecule that can have an amino acid attached to it?
   a. rRNA
   b. mRNA
   c. tRNA
   d. SnRNA
   e. All of the above

5. Which of the following is an RNA copy of part of one strand of DNA?
   a. rRNA
   b. mRNA

c. tRNA

d. SnRNA

e. All of the above

6. Aminoacyl-tRNA synthetase enzymes

   a. add amino acids to the growing polypeptide chain on the ribosome.

   b. are a part of the ribosome.

   c. add amino acids to tRNA molecules.

   d. include both a and b.

   e. include all of the above.

Short Answer

7. Compare ribosome assembly in the nucleolus with an automobile assembly plant.

Topic Test 2: Answers

1. **False.** Messenger RNA molecules are linked exons.

2. **True.** The small and large ribosomal subunits each contain both RNA and proteins.

3. **True.** These RNAs are parts of particles that cut introns out of pre-mRNA molecules.

4. **c.** tRNAs are shorter than mRNA and rRNA, and charged tRNA molecules contain an amino acid attached to one end.

5. **e.** All RNA molecules in cells are made from DNA using RNA polymerase, as discussed earlier in this chapter.

6. **c.**

7. Ribosome assembly is much more automatic. It is as if an automobile assembly plant could make cars by throwing engines, hoods, wheels, frames, and so on, into a room, give the room a few shakes to mix everything together, and out comes a bunch of fully assembled new cars!

TOPIC 3: THE GENETIC CODE

KEY POINTS

✓ *What is the genetic code? What are codons?*

✓ *What is redundancy? What is the molecular basis for redundancy?*

✓ *How universal is the genetic code among living organisms?*

The **genetic code** is like a dictionary that translates between the sequence of nucleotides in mRNA molecules and the sequence of amino acids in a polypeptide (protein). Each "word" is called a **codon**, and each is three letters long, with the four letters in the alphabet being A, U, G, and C, representing the four nucleotides found in mRNA. How many possible triplets are there? There are four possible letters in each of three positions, so there are 4 × 4 × 4, or 64 possible codons. Sixty-one of these code for one of the 20 amino acids. The remaining three are "stop" codons, which indicate that the end of a polypeptide chain has been reached.

The code is **redundant** since there is more than one codon for each of the amino acids. For example, there is more than one codon for phenylalanine, one of the amino acids. Both UUU and UUC specify phenylalanine. There are six codons that specify the amino acid leucine. By the middle 1960s, the entire genetic code had been worked out.

The code is almost totally **universal**; that is, the same code works for almost all living organisms. A few organisms have slight modifications of the code, but these are very rare.

The fact that a linear sequence of codons in the mRNA specifies a linear sequence of amino acids in a polypeptide is expressed by saying that the genetic code is **colinear**.

There is no punctuation or space between codons in the mRNA; one just follows the next. The triplet that is used to start most proteins, the **start codon**, is **AUG**. It normally specifies methionine, or a modified version of methionine. Any one of three triplets can specify the end of the chain: **UAA**, **UAG**, or **UGA**. These three **stop codons** were originally known as **nonsense codons**: If, through a mutation, they appeared in what was the middle of a protein, they would cause a premature termination of the growing chain. Thus, we had identified nonsense mutations before we developed the genetic code and learned of their normal role as stop codons.

Topic Test 3: The Genetic Code

True/False

1. The genetic code is colinear.

2. The genetic code translates between mRNA nucleotide sequences and amino acid sequences in proteins.

Multiple Choice

3. Concerning the genetic code, which of the following is correct?
   a. Plants and animals use different genetic codes.
   b. Each word in the genetic code is two letters long.
   c. There are 16 possible "words" in the genetic code.
   d. More than one "word" can specify the same amino acid.

4. The genetic code
   a. uses triplets of nucleotides to specify amino acids.
   b. is almost universal among living organisms.
   c. contains nucleotide "commas" or spaces, between each codon in the message.
   d. includes both a and b.
   e. includes all of the above.

5. Imagine that life is found on Mars, and that the DNA of Martian life contains only two nucleotides. Martian life-forms use 10 different kinds of amino acids to make proteins, and each of their codons are the same length. Each of their codons must be at least how many nucleotides long?
   a. One
   b. Two
   c. Three

d. Four

e. Five

Short Answer

6. Given an mRNA sequence, can one predict a unique corresponding amino acid sequence?

7. Given an amino acid sequence, can one predict a unique corresponding mRNA nucleotide sequence?

Topic Test 3: Answers

1. **True.** The linear sequence of triplet codons in RNA specifies the linear sequence of amino acids in a polypeptide.

2. **True.** Triplet nucleotides specify amino acids.

3. **d.** This is redundancy.

4. **d.** There are no "commas" or spaces between the codons.

5. **d.** One needs a minimum of 11 codons (one for each of the 10 amino acids plus one as a stop codon). If each codon were three nucleotides long, that would give $2 \times 2 \times 2$, or 8 codons (we are multiplying by two because there are only two nucleotides in the DNA, and so only two options at each position in the codon). With each codon being four nucleotides long, we get $2 \times 2 \times 2 \times 2$, or 16 codons, more than enough.

6. **Yes.** Each codon specifies a unique amino acid.

7. **No.** Each amino acid can be specified by several different codons, and so a number of different possible mRNA molecules could produce the same amino acid sequence.

TOPIC 4: TRANSLATION

KEY POINTS

✓ *What steps occur during translation?*

✓ *What are the roles of mRNA, tRNA, and ribosomes in translation?*

✓ *What repeating cycle of events occurs during the elongation phase of translation?*

Translation, or protein synthesis, involves the formation of chains of amino acids, linked by peptide bonds (Chapter 3). The formation of peptide bond linkages between amino acids is catalyzed by a ribozyme (part of the rRNA acting as a catalyst). If a single ribozyme does the linkage, why does the procedure need all of these players: ribosomes, mRNA, tRNA, aminoacyl-tRNA synthetases, and so on? All these players are needed because the cell does not link just any amino acids together. Cells must create specific sequences of amino acids in order to carry out cellular functions. The set of proteins in cells largely determines what the cell is and does. The difference between a human cell and a bacterial cell is in the proteins that are being made and what those proteins do. The difference in humans between a brain cell and a liver cell, again, is in the proteins made and what they do.

Here we consider the flow of information from RNA to protein—translation. There are three stages to translation: initiation, elongation, and termination. An analogy for the role of each of the players might help you to understand what is happening. The ribosome serves as a kind of a "tape recorder," with the mRNA serving as the tape. The tRNA molecules are something like the tape heads, reading and translating the message. The tRNAs are able to play this dual role because the anticodon on each tRNA serves as a kind of reader of the mRNA codon, while the amino acid attached to the tRNA becomes the translation. When the codon and anticodon "match," the amino acid on the tRNA is added to the growing polypeptide chain. We examine the details of the process below.

1. Initiation. The peptide chain begins to be made at the amino-terminal end; that is, there is a free amino group at the beginning end, and there will be a free carboxyl group at the other end of the polypeptide. There is a special **"initiator" tRNA-methionine** (methionine is its linked amino acid) that recognizes the AUG ("start") codon. In prokaryotes, a modified, formyl-methionine is the amino acid linked to the initiator tRNA. In all organisms, a small ribosomal subunit joins together with an mRNA and the initiator tRNA, and then a large ribosomal subunit joins to complete the initiation complex.

Some other, soluble proteins help in the process of protein synthesis. For the initiation phase these are called **initiation factors**. There also are **elongation factors** and even a **termination factor**. All are proteins that aid in the process of protein synthesis.

2. Elongation. Each round of elongation adds another amino acid to the chain. Elongation is repeated over and over, usually hundreds of times, to make each polypeptide. There are three substeps to elongation:

   a. Binding. The binding of the next tRNA is directed by the next codon on the mRNA at a site on the ribosome called the **A site (amino-acyl site)**. A GTP molecule is hydrolyzed as binding occurs, and the energy released probably is used in the recognition process, ensuring accuracy in the linkage between codon and anticodon.

   b. Peptide bond formation. The amino acid on the new tRNA is bonded to the amino acid chain attached to a tRNA molecule at the **P site (peptidyl site**; next to the A site on the ribosome). This step of peptide bond formation is catalyzed by a piece of rRNA. The fact that a **ribozyme** catalyzes peptide bond formation probably is of significance for the origin and evolution of life. The bond formation actually causes the growing polypeptide chain to be transferred to the new tRNA at the A site. The old tRNA, on the P site, now without any attached amino acids, leaves the ribosome. The P site now is temporarily vacant.

   c. Translocation. The tRNA, with growing polypeptide attached, shifts along the ribosome from the A site to the P site, and the mRNA shifts with it, moving one codon over and presenting the next codon on the mRNA at the now vacant A site. The vacant tRNA moves from the P site to a newly recognized E site (exit site) on the ribosome. The process of elongation now can repeat itself with the binding of a new, charged tRNA at the A site. During translocation a GTP molecule is hydrolyzed. Presumably, this gives energy for controlling the movement. Elongation is repeated, typically, hundreds of times as the polypeptide chain grows in length.

3. Termination. When a termination, or stop, codon (UAA, UAG, UGA) reaches the A site, a releasing factor (soluble protein) binds at the A site and triggers the breaking apart of the complex. The amino acid chain is split from the last tRNA, and the tRNA, mRNA, and small and large ribosomal subunits come apart, ending the process of translation with a new polypeptide having been formed.

Usually a single mRNA has a number of ribosomes running along it at any time, each making a polypeptide chain. The complex is called a **polyribosome**.

Each completed protein contains one or more polypeptide chains. Proteins themselves are degraded with time. A typical protein in a human cell has a half-life of about 2 weeks. Thus, protein synthesis always is going on, even in adults.

We have described the process of protein synthesis as if it were very organized, and in some ways it is. However, protein synthesis is highly dependent on **random collisions**, as are most processes in cells. At the ribosome, charged tRNAs randomly collide with ribosomes, as do many other ions and molecules in solution. Most frequently it is water molecules and ions that impact with the A site of the ribosome. Obviously, most such collisions are unproductive. Even most collisions with tRNA molecules will be with the wrong tRNA for the mRNA codon at the A site, or will be with uncharged tRNAs. Even collisions with the right tRNA can be in the wrong orientation. Nevertheless, the whole process of making a typical protein on a ribosome takes less than 1 minute. Obviously, many collisions must occur each second.

For information about **mutations and the genetic code**, see the web page.

Topic Test 4: Translation

True/False

1. Binding, peptide bond formation, and termination are the three substages that repeat during the elongation stage of protein synthesis.

2. Ribosomes contain A and P sites, and during elongation, charged tRNA molecules bind to the A site.

3. Peptide bond formation is catalyzed by an RNA molecule.

4. Anticodons are found on tRNA molecules.

Multiple Choice

5. During the elongation phase of translation,
   a. the stop codon indicates it is time to move on to the next codon.
   b. each new, charged tRNA binds to the P site on the ribosome.
   c. a translocation occurs and involves the simultaneous movement of both mRNA and tRNA with attached polypeptide relative to the surface of the ribosome.
   d. for a typical protein, four or five amino acids are put together to make one, complete chain.
   e. None of the above are correct.

6. A peptide bond links
   a. a carbon to a hydrogen.
   b. an oxygen to a carbon.
   c. a carbon to a nitrogen.

d. a carbon to a carbon.

e. an oxygen to a nitrogen.

Topic Test 4: Answers

1. **False.** The third substage is translocation, not termination.

2. **True.**

3. **True.** It used to be thought as a protein, but that assumption proved to be wrong.

4. **True.** Anticodons bind to mRNA codons during protein synthesis on the ribosome.

5. **c.** Stop codons indicate the end of the chain; new tRNAs bind to the A site; and hundreds of amino acids, not four or five, are linked.

6. **c.** If you missed this, you should go back and review the description of peptide bonds given in Chapter 3. Remember that a peptide bond links a carboxyl group (C) on one amino acid to an amino group (N) on the next.

APPLICATION: SICKLE-CELL DISEASE

An example of a point mutation is that which occurs in **sickle-cell disease**. A base-pair substitution in the DNA causes the production of mutant **hemoglobin**. Normal and mutant hemoglobin are compared below:

	Normal Hemoglobin	**Mutant Hemoglobin**
DNA sequence	CTT	CAT
Resulting mRNA	GAA	GUA
Amino acid in protein	Glu	Val

That single change in one amino acid out of 146 in the hemoglobin polypeptide changes the shape of the hemoglobin protein, and results in a change in shape of the entire red blood cell, compromising its ability to deliver oxygen to body tissues. This produces sickle-cell disease among those who have two copies of the mutant, sickle-cell gene. Tragically, while we have known the cause of sickle-cell disease for many years, we have not yet been able to find a cure for the disorder.

One puzzle that sickle-cell disease presented has been solved. In some tropical regions of Africa, almost one-third of individuals carry one copy of the mutant gene. The puzzle: If it is so detrimental, why is it such a common gene in certain human populations? Why hasn't natural selection eliminated, or at least greatly reduced, the frequency of this mutant gene? Individuals have two copies of the hemoglobin gene. As long as one copy is normal, individuals will not have the disorder (the sickle-cell gene is recessive—see Chapter 12). For full expression of the disorder, both copies must be the mutant, sickle-cell version. The reason why the mutant gene may be common is because individuals carrying one copy are resistant to malaria. Malaria is a serious disease that kills many who live in tropical regions. So, the benefit gained by those with one copy of the mutant gene must offset the disadvantage so obviously present in those who have two copies of the sickle-cell gene and develop sickle-cell disease.

Chapter Test

True/False

1. Codons consist of triplets of nucleotides in RNA.

2. During the elongation phase of translation, the three substeps are charged tRNA binding, peptide bond formation, and translocation.

3. Four different kinds of RNA are rRNA, tRNA, mRNA, and SnRNA.

Multiple Choice

4. RNA polymerase uses a DNA strand with the sequence GATTACA as a template for catalyzing the making of RNA. The complementary sequence of bases in the RNA is
   a. GATTACA.
   b. CTAATGT.
   c. CUAAUGU.
   d. AGCCGTG.
   e. TCGGCAC.

5. RNA polymerase catalyzes the synthesis of
   a. RNA from an RNA template.
   b. RNA from a DNA template.
   c. DNA from an RNA template.
   d. protein from an RNA template.
   e. protein from a DNA template.

6. Which of the following statements about ribosomes is false?
   a. Ribosomes contain protein.
   b. Ribosomes contain rRNA.
   c. Ribosomes contain DNA.
   d. Ribosomes contain two subunits.
   e. Ribosomes are involved in the synthesis of proteins.

7. The specifications for the sequence of amino acids in a protein must first be transcribed from DNA into
   a. protein.
   b. mRNA.
   c. tRNA.
   d. ribosomes.
   e. rRNA.

8. Imagine that aliens visit from another solar system. Their DNA contains six nucleotide bases, and their proteins are made from 30 amino acids. The minimum codon length for such aliens is
   a. one.
   b. two.
   c. three.
   d. four.
   e. five.

9. Concerning mRNA from eukaryotic cells, which of the following is false?
   a. An mRNA consists of linked exons.
   b. An mRNA specifies the sequence of amino acids in a polypeptide.
   c. Messenger RNA molecules have modified ends, such as poly-A tails.
   d. Messenger RNA molecules are found in the nucleus, but not in cytosol.
   e. A larger pre-mRNA has introns that are removed.

10. The genetic code
   a. is specific for different organisms. For instance, it is not the same for humans and birds.
   b. has only one triplet codon for each amino acid.
   c. is like a dictionary that translates from the sequence of bases in DNA to the sequence in RNA.
   d. will allow one to determine which amino acid will substitute for another, if a mutation has produced a known change in a known sequence of bases in a gene.
   e. allows one to identify a unique triplet of bases on an mRNA if one knows the amino acid present in the protein.

Chapter Test Answers

1. **T** 2. **T** 3. **T** 4. **c** 5. **b** 6. **c** 7. **b** 8. **b** 9. **d** 10. **d**

Check Your Performance:

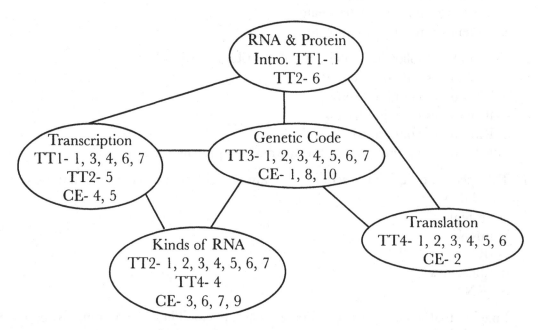

Key: TT = Topic Test; CE = Chapter Exam. Numbers indicate exam questions. Some questions are listed more than once if they refer to more than one topic.

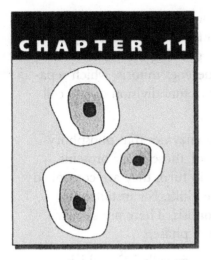

CHAPTER 11

Cell Cycle, Mitosis, and Meiosis

We know that all living organisms are made of cells, and that the cells divide. As cells divide, the genetic information in each parent cell is copied and distributed to the two daughter cells.

Cell (division) cycle, which includes **mitosis**, results in a separating of duplicate copies of the genetic material (doubling of the nucleus) prior to cell division. The cell cycle allows multicellular organisms to continue to grow by dividing their cells, while each cell is ensured a copy of the organism's genetic blueprints.

Another form of cell division is **meiosis**, which allows for the reduction in chromosome number that is necessary for sexual reproduction. Events during meiosis also contribute to genetic recombination, as we will see. **Genetic recombination** allows for a mixing of genes and thereby contributes to the genetic variation on which natural selection can act.

Chromosomes are the key to mitosis and meiosis. They contain most of the genetic material of the cell, as we learned in Chapter 5. Every species has a characteristic number of chromosomes. Human cells have 46 chromosomes; we inherit half from our mothers and half from our fathers. It is meiosis that reduces the number to half in germ-line cells, so that sexual recombination can take place.

ESSENTIAL BACKGROUND

- Chromosomes (Chapter 5)
- Microtubules (Chapter 5)

TOPIC 1: THE CELL CYCLE AND MITOSIS

KEY POINTS

✓ *What steps occur during the cell cycle?*

✓ *What happens to the DNA in chromosomes during interphase?*

✓ *How is the cell cycle controlled? What are cyclins and cyclin-dependent kinases?*

✓ *What are the stages of mitosis, and what happens to chromosomes?*

✓ *What is cytokinesis?*

The **cell cycle** (cell division cycle) describes the process that culminates in cell division. It is a cycle that repeats itself over and over as a fertilized egg becomes a multicellular organism. The cell cycle has two parts: **interphase** and **mitosis**. Most of the time a cell is in interphase, which consists of three phases: an initial growth phase (**G₁**), a phase of DNA synthesis (**S**), and a phase of further growth (**G₂**) before cell division. The cell usually is growing during all three parts of interphase. At the end of interphase (end of G_2), the cell undergoes mitosis, which separates the cell's duplicated chromosomes, and finally, **cytokinesis**, the actual division of the cell into two daughter cells.

The phases of the cell cycle are tightly controlled and orchestrated by enzymes that phosphorylate (add phosphate groups to) other proteins, controlling the function of those other proteins. These enzymes are called **cyclin-dependent kinases (Cdks)** and, in turn, they are controlled by **cyclins**, which are proteins that attach to, and usually activate, the Cdks. For instance, a cyclin and a Cdk control the exit from interphase and beginning of mitosis. There are similar controls that act at other switching points, such as that from G_1 to the S phase.

Another set of controls operate when cells cease dividing and enter a stage called **G_0**, which blocks the continuation of the cell cycle. Cells usually exit the cycle during the G_1 phase and enter G_0 before DNA replication. Some cells, such as muscle cells and neurons, will never again enter the cell cycle, but other cells in G_0 can resume cell cycling if induced by special growth factors or tissue damage. For instance, damage to an adult's liver can trigger liver cells to begin to duplicate and grow a replacement for the damaged part of the liver.

Mitosis, which starts during the cell cycle at the end of G_2, has its own set of stages: **prophase**, **prometaphase**, **metaphase**, **anaphase**, and **telophase**. These stages involve the separation of the cell's chromosomes, which is accomplished in such a way that a complete copy of all of the DNA, which was duplicated during the S phase of interphase, gets to each daughter cell. Each chromosome has one long strand of DNA in it, and during most of the life of the cell the DNA strand in the chromosome is spread out. Proteins are associated with the DNA, and the complex is referred to as **chromatin**.

The chromosomes condense during prophase, and each condensed chromosome can be seen to consist of pairs of **sister chromatids**. These contain the duplicated DNA molecules, one each. All of the chromosomes align in a plane (**metaphase plate**) in the middle of the cell during metaphase, and separate into independent chromosomes during anaphase, when the divided pairs of what were sister chromatids head to opposite sides of the cell, allowing each daughter cell to gain a copy of each chromosome.

The alignment and movement of the chromosomes during mitosis are governed by the **mitotic spindle apparatus**, which contains microtubules (Chapter 5). These attach to the chromosomes and during anaphase, are responsible for pulling the paired chromosomes apart, toward the centrioles at either end of the cell. Nuclei begin to form around each complete set.

Cytokinesis completes cell division by forming a membrane that divides the cell into two, each containing one set of chromosomes. In animal cells, microfilaments (Chapter 5) are involved in producing a pinching of the cell's membrane that forms a **cleavage furrow**. The cleavage furrow continues to narrow until two cells are pinched apart in the middle. In plant cells, membrane-bound vesicles from the Golgi apparatus (Chapter 5) align in the middle of the cell and through fusion, form a **cell plate**, which becomes a new cell wall separating the two new daughter cells.

If all has gone well, the two daughter cells have identical genotypes—the same genetic material as the parent cell had.

Topic Test 1: Cell Cycle and Mitosis

True/False

1. Cytokinesis is the actual division of a parent cell into two daughter cells—a splitting of the cell into two.

2. Mitosis is the process of duplicating the DNA in chromosomes prior to cell division.

3. The mitotic spindle apparatus contains microtubules.

Multiple Choice

4. During metaphase,
   a. chromosomes condense.
   b. the condensed chromosomes are aligned in a plane in the middle of the cell.
   c. sister chromatids separate and head for opposite poles of the cell.
   d. daughter nuclei form and a cleavage furrow forms.

5. DNA replication takes place during
   a. metaphase.
   b. telophase.
   c. anaphase.
   d. prophase.
   e. none of the above.

Short Answer

6. During one cell cycle,
   a. how many times is the DNA copied?
   b. how many times does the cell divide?
   c. how many daughter cells are formed?
   d. do the daughter cells contain identical genetic information?

Topic Test 1: Answers

1. **True.** During cytokinesis, membrane is added or pinched in such a way as to produce two cells from one.

2. **False.** DNA is duplicated during the S phase of interphase, not during mitosis itself.

3. **True.** The microtubules play a primary role in the separation of chromosomes that occurs during mitosis.

4. **b.** The other events occur before or after metaphase.

5. **e.** DNA replication takes place during the S phase of interphase, not during the listed stages of mitosis.

6. **a.** Once   **b.** Once.   **c.** Two   **d.** Yes, unless a mutation has occurred

TOPIC 2: MEIOSIS

KEY POINTS

✓ *What kind of cells does meiosis occur in?*

✓ *What does meiosis accomplish?*

✓ *How many cell divisions are involved, and what happens to the chromosomes during those cell divisions?*

✓ *How does meiosis compare to mitosis?*

Sex makes life more interesting and varied, but also makes it more complicated. **Sexual reproduction** introduces variety by mixing combinations of genes from two parents. This contrasts with **asexual reproduction**, such as occurs in bacteria, where the two daughter cells have the same genetic material, ignoring mutations. Populations of organisms that have the same genetic makeup are called **clones**. Bacteria form clones, and so can some plants, such as carrots, that reproduce by growing adult plants from parts of the parent.

Most human cells have 46 chromosomes, consisting of **23 pairs**. The two chromosomes in each pair are called **homologous chromosomes**, or homologues. One of each pair comes from the father, and the other from the mother. Homologous chromosomes usually contain the same genes (although the gene pairs are not always exactly identical—one might be a mutant, or variant, of the other). Thus, we get two copies of most of our genes, one from each parent. The main exception concerns one of the 23 pairs, the **sex chromosomes X and Y**. This pair does not contain identical genes. Human females have two X chromosomes and males have an X and a Y chromosome. The other 22 pairs of chromosomes are called **autosomes**.

Sexual reproduction involves reproductive cells called **gametes**. In humans these are **sperm cells** (in males) and **egg cells**, or **ova** (in females). Gametes contain half the number of chromosomes that other cells have. Meiosis is the key to making sperm and egg cells with half the number of chromosomes that are normally present in other cells. Such cells are said to contain a haploid (n) number of chromosomes—a single set. The other cells are said to be diploid (2n).

Combining the chromosomes from two haploid gametes results in the diploid number of chromosomes, which is what **fertilization** is all about. In contrast, getting from a diploid number of chromosomes to a haploid number is what meiosis is all about.

Meiosis consists of one doubling of the DNA in a diploid cell followed by two rounds of cell division. We end up with four haploid cells. The two rounds of cell division are called **meiosis I** and **meiosis II**.

Before **meiosis I**, the cell goes through a normal interphase, including a doubling of the DNA. In prophase I, there already are differences from mitosis. The condensed chromosomes are clustered in **tetrads**, containing pairs of homologous chromosomes, each with pairs of sister chromatids, for a total of four chromatids per tetrad.

During metaphase I, the homologous chromosomes line up in pairs (remain as tetrads), and instead of sister chromatids separating during anaphase I, the homologous chromosomes split and end up in separate daughter cells as meiosis I is completed. After cytokinesis, we have two cells with a different genetic makeup. Each cell has a haploid set of chromosomes, but each chromosome still consists of sister chromatids.

Meiosis II starts quickly, without interphase and without any further duplication of DNA. The sister chromatids separate during anaphase II, becoming individual chromosomes, and cytokinesis then produces a total of four haploid daughter cells.

We can compare mitosis and meiosis to gain an idea of the differences:

	Mitosis	Meiosis
Number of rounds of DNA replication	1	1
Number of rounds of cell division	1	2
Number of daughter cells	2	4
Number of chromosomes in each	Diploid (2n)	Haploid (n)
Genetic makeup of daughter cells	Same	Different
Role	Development	Reproduction

Topic Test 2: Meiosis

True/False

1. Homologous chromosomes refer to the pairs of similar chromosomes that were inherited, one from each parent.

2. Tetrads, each containing homologous chromosome pairs, form during meiosis II.

3. Each tetrad contains four sister chromatids in a homologous pair of chromosomes.

Multiple Choice

4. Comparing mitosis and meiosis,
   a. both consist of two cell divisions.
   b. both generate daughter cells that contain identical genetic material.
   c. tetrads form in both.
   d. daughter cells after mitosis contain the same amount of DNA as do daughter cells at the end of meiosis.
   e. None of the above is correct.

5. In both mitosis and meiosis,
   a. sister chromatids are present.
   b. tetrads form.
   c. a single round of DNA replication is followed by two sets of cell divisions.
   d. each daughter cell contains a haploid number of chromosomes.
   e. all of the above are correct.

6. In a nonfertilized egg cell (ovum) of a human female,
   a. there are 23 chromosomes, each containing a single, long DNA molecule.
   b. there are 46 chromosomes, each of which contains many DNA molecules.
   c. there are both X and Y chromosomes.
   d. there can be one Y chromosome.
   e. both c and d are correct.

Short Answer

7. What does meiosis do that mitosis cannot do?

8. When do sister chromatids separate during meiosis, and when do they separate during mitosis?

Topic Test 2: Answers

1. **True.** We inherit one of each homologous pair from each parent.

2. **False.** Such tetrads form and separate during meiosis I.

3. **True.** Often these exhibit crossing-over of material between the homologous pairs.

4. **e.** Incorrect answers a, b, and c each describe what is true of one but not the other.

5. **a.** Incorrect answers b, c, and d each describe what is true of one, but not the other.

6. **a.** Unfertilized human ova cannot contain Y chromosomes, since these are found only in males.

7. Meiosis produces a reduction in chromosome number, from diploid to haploid, which sets the stage for sexual reproduction.

8. During mitosis, sister chromatids separate during anaphase. In meiosis, sister chromatids separate during anaphase II.

TOPIC 3: SEXUAL REPRODUCTION AND GENETIC VARIATION

KEY POINTS

✓ *What are three sources of genetic variation during sexual reproduction?*

✓ *What is crossing-over, and how does it contribute to genetic variation?*

In sexual reproduction, three mechanisms contribute to genetic variation in offspring: independent assortment of chromosomes, crossing-over, and random fertilization.

Independent Assortment of Chromosomes During Meiosis

There is an **independent assortment of chromosomes** during meiosis I. **Tetrads**, containing homologous pairs of chromosomes, align along the metaphase plate during meiosis I. The orientation of the pairs is random; that is, if one of the tetrads is oriented with the paternal chromosome (chromosome that has come from the father of the individual) on the left, the neighboring tetrad might have either the paternal or the maternal chromosome oriented in the same direction, with a 50–50 chance. Thus, when the chromosomes within each tetrad separate, there is a random assortment of the pairs. The direction of sorting of one paternal and maternal homologous pair during cell division is independent of the other pairs. This is Mendel's rule of independent assortment, and will be considered in detail in the next chapter. Each daughter cell gets one of each kind of chromosome, but the distribution of paternal and maternal versions is random, like the flipping of a coin for each pair.

We can calculate the number of combinations of paternal and maternal chromosomes possible from such random assortment. The number goes up quickly as the haploid number of chromosomes in an organism increases. In an organism with a haploid number of 2 (n = 2), there are 2^2, or four different combinations, like flipping a coin twice: mom-mom, mom-dad, dad-mom, and dad-dad. In humans, with a haploid number of 23 (n = 23), there are 2^{23}, or about 8 million different combinations. Thus, random assortment alone allows each of us to produce up to 8 million genetically different gametes. That's a lot of variety, but it is just the beginning.

Crossing-over

Within a tetrad, which is formed during meiosis I, typically there are several places where the homologues have exchanged parts of chromosomes with each other. We refer to this as **crossing-over**, and to the site of crossing-over as a **chiasma**. There is an exchange of genetic material, and this is a form of genetic recombination, a new mixing of genes from the two chromosomes. Here, we are combining genes from maternal and paternal chromosomes. There is a precise splitting of the DNA in one chromatid of each of the homologous chromosomes, and a cross-splicing of that DNA. It has to be precise to avoid introducing mutations (due to insertion and deletion) that would result if the cutting were not at the same site on the two chromatids. During meiosis, each human chromosome undergoes about three such crossing-over events. This form of genetic recombination yields a great amount of variety, as the crossing-over can occur almost anywhere along the chromosome.

Random Fertilization

Random fertilization contributes a third source of variety. Even ignoring the variety introduced by crossing-over, each human partner contributes any one of 8 million different combinations of paternal and maternal chromosomes. With two parents, there are trillions of different possible genetic outcomes (8,000,000 × 8,000,000 = 16 trillion) as random sperm fertilizes random ovum.

With this genetic variety in offspring there is little wonder that siblings can be so different from one another. All of the variety is important for evolution. Favorable variations and combinations of genes will accumulate through natural selection (see Chapter 16). The diversity generated by sexual reproduction also ensures a source of genetic differences within populations that enable adjustments to be made for changes in environments. Mutations give us the genetic variation in the first place; sexual reproduction is a way of mixing and matching that has proved to be advantageous.

Topic Test 3: Sexual Reproduction and Genetic Variation

True/False

1. Crossing-over refers to the exchange of parts of chromatids in a tetrad with one another.

2. The cutting point for crossing-over can occur almost anywhere along a chromosome.

3. Crossing-over occurs between two different tetrads during meiosis.

Multiple Choice

4. Crossing-over
   a. involves the exchange of genetic material.
   b. contributes to genetic variation.
   c. occurs during meiosis.
   d. is an exchange of material between homologues.
   e. involves all of the above.

5. Sources of genetic variation in sexually reproducing organisms include
   a. independent assortment of chromosomes during meiosis.
   b. crossing-over seen in tetrads during meiosis.
   c. random fertilization.
   d. both a and b.
   e. all of the above.

Short Answer

6. Considering only independent assortment, how many genetically different gametes would an organism with three (haploid) chromosomes be able to produce?

Topic Test 3: Answers

1. **True.** This is one form of genetic recombination.

2. **True.** This allows crossing-over to introduce a considerable amount of variation.

3. **False.** Crossing-over occurs within a single tetrad.

4. **e.** These help to define crossing-over.

5. **e.** These are the three sources that were described earlier.

6. Eight (there are three chromosome pairs and each has two possible ways of being divided, giving $2^3 = 8$ possibilities. Another way to see this is to consider the three chromosome pairs as Aa, Bb, and Cc. The eight combinations are ABC, AbC, ABc, Abc, aBC, abC, aBc, and abc.

APPLICATION: CANCER

One possible result of a disruption in control of the cell cycle is **cancer**. Regulatory molecules and pathways that activate the cell cycle sometimes are induced inappropriately, or are not properly suppressed. We can see the result of this by comparing the growth of normal and cancer cells in culture dishes. Normal cells in cell culture on petri dishes will stop growing once they have covered the dish with a single layer of cells. This is called **density-dependent inhibition**. Cancer cells do not exhibit this control and continue to grow on top of one another on the dish. We speak of the cancer cells as being **transformed**.

Some of the pathways and molecules that can cause cancer have positive roles to play in the control of cell division, but can get out of hand owing to mutation. Some of the pathways are normally triggered by hormones or growth factors, and normally are suppressed without such inducers. Such growth factors come from outside of cells and usually bind to protein receptors on the surface of cells, inducing a series of events in the cell. One cause of cancer is for the cancer cell to have the pathway disrupted in such a way that the signals no longer are necessary, or may even inappropriately generate their own growth factors. Some cancers can be caused by viruses, and some are inherited, but the majority are caused by mutations inside our body (somal) cells that occur during our lifetime. These mutations can be caused by a range of factors including carcinogens in the food we eat, sunlight causing skin cancer, and smoking causing lung cancer. Some of the mutations occur in so-called **oncogenes**, altering the gene products and inducing cancer. Mutation also can block the action of other, so-called **tumor-suppressor genes**. One such tumor-suppressor gene, p53, commonly is mutated in lung cancer. The protein product of p53 normally blocks cell cycling during G_1.

The immune system often will destroy cancer cells, but sometimes seems not to recognize particular cancer cells as abnormal. The loss of control over cell division in such cells then results in continued, unchecked growth of cancerous cells, forming a **tumor**—a lump of growing cells. Such tumors can be benign and not cause serious problems because they do not spread.

Malignant tumors can spread to other body tissues, a process known as metastasis. These can be deadly and are usually treated with chemotherapy or radiation therapy. Searches for better cures for cancer continue.

Chapter Test

True/False

1. Chromosomes are separated for distribution to daughter cells during mitosis and meiosis.

2. Mitosis occurs during development, as the fertilized egg becomes a multicellular adult.

3. Crossing-over occurs during mitosis.

Multiple Choice

4. Concerning meiosis, which of the following is NOT correct?
   a. Tetrads form during meiosis I.
   b. There are two rounds of cell division during meiosis.
   c. Crossing-over occurs during meiosis I.
   d. Each daughter cell gets the same genetic information.
   e. Chromosome number is reduced from diploid to haploid during meiosis.

5. Given an organism that has 10 pairs of homologous chromosomes, and ignoring crossing-over, how many possible different combinations of chromosomes will occur during gamete formation?
   a. 10
   b. 20

c. 10^2
d. 2^{10}
e. 5

Short Answer

6. What is independent assortment?

Chapter Test Answers

1. **T**   2. **T**   3. **F**   4. **d**   5. **d**

6. Independent assortment is the random separation of homologous chromosomes that occurs during meiosis.

Check Your Performance:

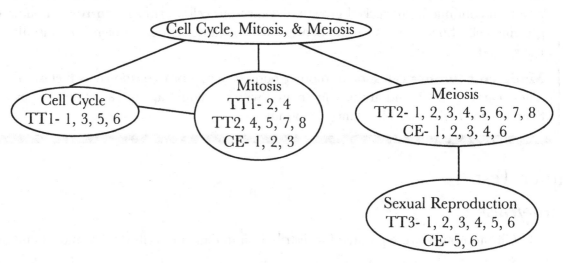

Key: TT = Topic Test; CE = Chapter Exam. Numbers indicate exam questions. Some questions are listed more than once if they refer to more than one topic.

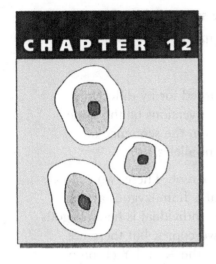

CHAPTER 12

Mendelian Genetics

In the mid-1800s, at about the time when Darwin was establishing evidence for evolution and natural selection, a monk named **Gregor Mendel** studied the inheritance of certain traits in garden peas. Mendel's work was not recognized as important until about 1900, after his death. He now has a whole field of study named after him. In this chapter and the next, we examine the rules that he uncovered, develop an understanding of their basis, and then add to them, refine them, and examine some exceptions to them. You will see that Mendel's rules, and even the exceptions to them, can be predicted or understood from the information about genes in Chapters 9 and 10, and the information about chromosome separations during meiosis in Chapter 11.

ESSENTIAL BACKGROUND

- Meiosis and chromosomes (Chapter 11)
- Genes and the nature of genetic information (Chapters 9 and 10)

TOPIC 1: MENDEL'S RULES AND GENETIC VARIATION

KEY POINTS

✓ *What are traits, and what are alleles?*

✓ *What do dominant and recessive mean?*

✓ *What is Mendel's rule, or law, of segregation?*

✓ *What is Mendel's rule of independent assortment, and what causes exceptions to this rule?*

We begin with some essential vocabulary:

Character: a heritable feature, such as eye color.

Trait: a **variant** for a character, such as blue eyes.

Allele: genetic variant at a given gene locus (site on the chromosomes). Alleles have different DNA sequences, and were initially created by mutations.

Genotype: the genetic makeup of an individual.

Phenotype: the appearance, structure, function, or behavior of the individual. Phenotypes are caused by a combination of genotype plus environment, and the interactions between the two.

Homozygous: the two copies of a gene in an individual are the same allele.

Heterozygous: the two copies of a gene in an individual are different alleles.

Now we can add a few observations, or rules:

1. Different traits, or variations in inherited characters, are accounted for by different versions, or alleles, of genes. The concept of alleles, of different versions of the same gene, is a critical one to grasp. Alleles are caused by differences in the sequence of nucleotide bases in DNA, and mutations produce these different alleles.

2. An organism inherits two copies of each gene, one from each parent. These two copies can be the same allele, in which case we say that the individual is homozygous at that gene locus. If the two copies of the gene differ, we say that the individual is heterozygous at that gene locus. (Most genes in most organisms consist of two copies, but there are exceptions to this rule. One obvious exception are the genes on the X and Y chromosomes in human males—each male has only one copy of each of these genes, since the X and Y chromosomes do not share any genes in common. We will consider this complication in the next chapter.)

3. If the two gene copies in an individual differ at a given gene locus (i.e., if the individual is heterozygous), there can be a **dominant** allele that appears to be fully expressed, and a **recessive** allele that has no obvious effects on phenotype. (There are exceptions to this as we learn later in this chapter, and even the question of whether the recessive allele is expressed can depend on the level at which one looks.)

4. The two copies of each gene segregate during gamete production (meiosis, discussed in Chapter 11). Each gamete (ovum or sperm) contains only one of the two genes for each character. This is called **Mendel's rule** (or **principle**, or **law**) **of segregation**. (Rare exceptions to this rule can occur and can produce individuals with genetic defects, such as Down syndrome, which we will discuss in Chapter 15.)

5. The genes for different characters sort independently during gamete production. This is **Mendel's rule of independent assortment**, which was described in detail for individual chromosomes in Chapter 11. (There are exceptions to this independent assortment rule for genes because some genes are linked together on the same chromosome, and so do not sort independently. We will examine linkage in the next chapter.)

Topic Test 1: Variation Mendel's Rules and Genetic

True/False

1. The genotype of an individual is produced by a combination of phenotype plus environment.

2. Living organisms inherit one copy of each of their genes.

3. Recessive alleles are expressed in heterozygotes.

4. Mendel's rule of independent assortment implies that if a particular pair of genes segregates into gametes such that the allele from an individual's mother goes to one

gamete, one cannot tell, for another gene, whether the allele from the mother will go to the same gamete.

5. A character can have more than one trait.

Multiple Choice

6. At a particular gene locus, there are two possible versions, or alleles, S and s. When the two are present in an individual, only the S allele appears to be expressed. Which of the following would be the correct way to describe an individual whose genotype is ss?
   a. Homozygous dominant
   b. Homozygous recessive
   c. Heterozygous

7. Mendel's rule of segregation, or its equivalent, states that
   a. the two copies of each gene separate during gamete formation.
   b. one inherits only one copy of each gene from each parent.
   c. offspring can be heterozygous, but not homozygous.
   d. offspring can be homozygous, but not heterozygous.
   e. both a and b are correct.

Topic Test 1: Answers

1. **False.** The phenotype is produced by a combination of genotype plus environment.

2. **False.** In most organisms, and for most genes, two copies are inherited.

3. **False.** The dominant allele will be expressed in a heterozygote.

4. **True.** Mendel's rule states that the sorting of the different genes is random, or independent.

5. **True.** Traits are different versions of a character, such as different colors for the flower of a plant.

6. **b.** Since Ss individuals have an "S" phenotype, the "s" allele must be recessive. Then, ss individuals would be homozygous for the recessive allele, or homozygous recessive.

7. **e.** These are two ways of saying the same thing. Concerning choices c and d, offspring can be heterozygous or homozygous, depending on the genotypes of the parents.

TOPIC 2: GENETIC CROSSES

KEY POINTS

✓ *How does one determine the nature of offspring and their frequencies from monohybrid and dihybrid crosses?*

✓ *What is a test cross?*

Here is more essential vocabulary:

P: parental generation.

F1: first-generation offspring (F stands for filial, offspring).

F2: second-generation offspring, generally obtained by crossing F1 progeny (brothers, sisters) with each other.

Monohybrid cross: a genetic cross, or mating, that follows the inheritance of one character.

Dihybrid cross: a genetic cross, or mating, that follows the inheritance of two characters.

The Monohybrid Cross

Presume that one character for a plant is the color of its flower. If the dominant allele, B, for flower color gives black flowers, and the recessive allele, b, gives white flowers, then a plant with homozygous dominant alleles will have black flowers and the genotype BB, while a plant with homozygous recessive alleles will have white flowers and the genotype bb. Heterozygous plants are Bb in genotype, and will have black flowers since B is the dominant allele. Usually, the dominant allele is indicated by a capital letter, while the recessive allele is indicated by a lowercase letter.

Let us consider a simple **monohybrid cross** between two, parental-generation individuals with genotypes BB and bb. To determine the nature of the progeny, or offspring, or F1 hybrids, of the cross, it is necessary to consider all of the possible gametes from the parents. In this case, that is easy because the homozygous dominant individual has a genotype BB, and so all of its gametes have the B allele. The homozygous recessive individual has a genotype bb, and so all of its gametes have the b allele. All of the F1-generation offspring combine the alleles from the two gametes, or B + b = Bb for all of the F1-generation progeny. Thus, all of the F1 individuals are heterozygous, Bb, and since B is dominant, all have black flowers.

Now let us consider the F2 generation, obtained by crossing two F1 heterozygotes with one another. This situation is a bit more complex, since each F1 individual can produce two possible gametes, either B or b, with equal probability. We set up a little square showing the gamete options (in **bold**) on the outside, and then combine the gamete types to produce the possible F2-generation individuals:

	Female Gamete Types	
Male Gamete Types	**B**	**b**
B	BB	Bb
b	bB	bb

These four possible offspring, BB, Bb, bB, and bb, are in roughly equal abundance, a probability of one chance in four for each, since the two types of gametes from each F1 individual are equally abundant. In terms of their genotypes, the offspring are of three types—homozygous dominant (BB), heterozygous (Bb and bB), and homozygous recessive (bb)—and these three types are found in an abundance ratio of 1:2:1, respectively.

If one examines phenotypes for this F2 generation, three (BB, Bb, and bB) of the four plants will have black flowers (dominant) and one (bb) of the four will have white flowers. Thus, the ratio of phenotypes black to white is 3:1. Notice the difference in ratios, depending on how one expresses them. Also realize that Bb and bB are genetically equivalent—both are heterozygous.

The Test Cross

Consider a plant whose phenotype is black flowers. Its genotype could be BB or Bb, since black is dominant. To determine whether the genotype is BB or Bb, we do what is called a **test cross**: Cross the unknown genotype (BB or Bb) with a homozygous recessive individual (bb). There are two possibilities: If the unknown genotype is BB, then the result would be the same as the monohybrid cross discussed earlier, an F1 generation all of whose offspring have black flowers. If the unknown genotype is Bb, then the offspring will have the following genotypes:

	Female Gamete Types	
Male Gamete types	**b**	**b**
B	Bb	Bb
b	bb	bb

Half the offspring have a genotype of Bb, and so a phenotype of black flowers, while the other half of the offspring have a genotype of bb, and so a phenotype of white flowers.

The Dihybrid Cross

In a **dihybrid cross**, one is following the inheritance of two different characters, with two traits for each one. Let us consider the two characters to be seed color, with green and yellow as traits, and seed shape, with round and wrinkled as traits. For color we have the alleles Y = yellow (dominant) and y = green (recessive), and for shape we have R = round (dominant) and r = wrinkled (recessive). Consider the following cross: YYRR × yyrr. The first parent is homozygous dominant for each character, and has a phenotype that is yellow-round. The second parent is homozygous recessive for each character, and has a phenotype that is green-wrinkled. We begin our analysis of the cross by determining the possible gametes from each parent. There is but one possibility for each, YR and yr, respectively. The F1-generation offspring must be the combination of these gametes, or YR + yr = YyRr. All of the offspring are heterozygous for both color and shape of seeds, and all have the same phenotype, yellow-round. For the F2 generation we cross F1 with F1: YyRr × YyRr. Now there are four possible gametes for each parent: YR, Yr, yR, and yr. Notice that each of these four were obtained by picking one of the possible alleles at each of the gene loci, either Y or y at one, and either R or r at the other. Notice that all four gametes are equally likely, one chance in four for each, as long as independent assortment of these two genes is occurring. We can set up another square, such as we did with the monohybrid cross earlier, with the possible gametes (in **bold**) on the outside and the F2-generation offspring in the middle. Now there are more possibilities:

	YR	**Yr**	**yR**	**yr**
YR	YYRR	YYRr	YyRR	YyRr
Yr	YYrR	YYrr	YyrR	Yyrr
yR	yYRR	yYRr	yyRR	yyRr
yr	yYrR	yYrr	yyrR	yyrr

This is called a **Punnett square**. Since the odds of each gamete forming is equal, at one in four, the probability of each F2-generation outcome is one in 16. The 16 F2-generation outcomes can be grouped into four phenotypes:

yellow-round   9   (YYRR, YYRr, YyRR, YyRr, YYrR, YyrR, yYRR, yYrR, yYrR)
green-round   3   (yyRR, yyRr, yyrR)
yellow-wrinkled   3   (YYrr, Yyrr, yYrr)
green-wrinkled   1   (yyRR).

Instead of just memorizing the 9:3:3:1 ratio of the four phenotypes, you should understand how to do these dihybrid crosses more generally, by determining what gametes are possible for each genotype, and then forming the Punnett square, or an equivalent, to determine the possible outcomes given the gametes possible from each parent. There are many other possible dihybrid crosses, such as YyRR × yyRr, and they do not result in the 9:3:3:1 ratio. You can develop your skills here by working the problems.

Topic Test 2: Genetic Crosses

True/False

1. Two individuals with genotypes Bb and bB are genetically identical.

2. In the dihybrid cross shown earlier, there were more dominant than recessive traits among the offspring because it only takes one copy of the dominant gene to produce the dominant phenotype, whereas both copies must be recessive in order to have the recessive phenotype.

Multiple Choice

3. Consider the following monohybrid cross (A = dominant, a = recessive): aA × aa. The fraction of progeny that are
   a. homozygous recessive is 50%.
   b. heterozygous is 25%.
   c. homozygous dominant is 50%.
   d. homozygous recessive or homozygous dominant is 100%.
   e. None of the above is correct.

4. Consider the following cross: AaBB × Aabb. Among the F1 generation of offspring, what percentage are heterozygous at both gene loci?
   a. 0%
   b. 25%
   c. 50%
   d. 75%
   e. 100%

Short Answer

5. Consider the following cross: YYrr × YyRR, where Y (yellow color) is dominant over y (green), and R (round shape) is dominant over r (wrinkled). In the F1 generation, assuming independent assortment,
   a. what are the genotypes and their ratios?
   b. what are the phenotypes and their ratios?

Topic Test 2: Answers

1. **True.** At this gene locus, at least, the two are the same. The order of the genes does not change anything.

2. **True.** Heterozygotes express the dominant phenotype.

3. **a.** Setting up the cross involves determining the possible gametes from each parent. We have a or A from one, and only a from the other, so we can set up a simplified table:

	a
A	Aa
a	aa

Half of the offspring will be heterozygous (Aa), and the other half will be homozygous recessive (aa). None will be homozygous dominant (AA).

4. **c.** Setting up the cross, we first determine the possible gametes, which are AB and aB from the first parent, and Ab plus ab for the second, so we can set up another simplified table:

	Ab	**ab**
AB	AABb	AaBb
aB	aABb	aaBb

Two of the four are heterozygous at both gene loci (AaBb, aABb), while the other two are homozygous at one of the two gene loci (AABb, aaBb).

5. To answer the question, we must begin by determining the possible gametes. For YYrr, only one gamete is possible, Yr. For YyRR we have two possible gametes, YR and yR. So the cross is:

	YR	**yR**
Yr	YYrR	YyrR

There are two genotypes, as shown, and they are in equal abundance, 1:1. There is one phenotype, yellow-round, since each F1 progeny has at least one dominant gene at each loci.

TOPIC 3: COMPLEXITIES

KEY POINTS

✓ *What are incomplete dominance and codominance?*

✓ *What are multiple alleles?*

✓ *What is pleiotropy?*

✓ *What is epistasis?*

✓ *What is polygenic inheritance?*

There are some complicating features to the simple kind of inheritance described in the previous section. When we have two alleles for a character, one is not always dominant while the other is recessive. Sometimes there is **incomplete dominance**, and the phenotype appears to be a blend of the two traits. However, it is important to realize that the genes themselves have not blended; they can be separated by further crosses. Snapdragon color is an example of incomplete dominance. A red allele produces red flowers when it is homozygous, and a white allele produces white flowers when homozygous. When the two are together (heterozygous), a pink flower is produced. A pink-flowered snapdragon, when crossed with itself, will produce F1 progeny that are in the following ratios: one of four with red, one of four with white, and two of four with pink flowers. You should do the cross to check that claim.

Another possibility is **codominance**, when both phenotypes are present in the setting of heterozygous alleles. An example of this is blood types, which include A and B. These blood types refer to molecules present on the surface of red blood cells. Individuals who have one A allele and one B allele are AB in their genotype and also show both A and B molecules on the surface of their red blood cells. This presence of both types of molecules can be detected with special antibodies to A and B, and is a phenotype.

We have simplified this discussion on genetics by looking at a maximum of two different alleles at a given gene locus. Although individuals are limited to two, a population of individuals can have more than two alleles for a given gene locus. Then we speak of **multiple alleles**. We can return to blood types for a simple example of multiple alleles. There actually are three blood types, A, B, and O. Individuals with blood type O have no glycoprotein on the surface of the red blood cell, and so represent a third alternative to glycoproteins A and B. Individuals then can have a blood genotype of AO, AB, BO, OO, AA, or BB.

Another effect is called **pleiotropy**, which refers to genes that can affect more than one character. A gene may make just one polypeptide, but that polypeptide can affect a number of different properties of an organism. The hemoglobin gene, described in Chapter 10 when sickle-cell disease was discussed, is an example of a pleiotropic gene. Sickled cells cause a breakdown of red blood cells and a higher likelihood of red blood cell clumping, which can block blood vessels and cause impaired mental function and kidney failure, and increase the risk of pneumonia. These are different phenotypic characteristics caused by a defect in a single, pleiotropic gene.

Epistasis is the ability of one gene to influence the expression of another gene, or even a group of genes. The existence of epistasis should not be surprising if one considers biochemical pathways. Several different genes are involved in such pathways, and if an early step in the pathway is blocked, then the precursor for a later gene is not made, and thereby the action of the later gene is disrupted. Expressions of alleles at that later point in the pathway can be blocked by epistasis.

Most characters are not inherited in the simple way that Mendel described. Most characters depend on a range of genes and the interactions of their products. Such **multigenic**, or **polygenic**, characters in humans would include intelligence, height, and skin color. Furthermore, most characters are the result not just of genotype but also of environment. In many cases we have polygenic inheritance that is influenced in complex ways by environmental factors and experiences. An organism's phenotype is the consequence of its genotype plus its unique environmental history. Each of us is a unique individual.

Topic Test 3: Complexities

True/False

1. Pleiotropy refers to the fact that most characters depend on more than one gene.

2. Incomplete dominance is the same as codominance.

3. Two alleles for eye color are red and white. A heterozygous individual has pink eyes. This is an example of incomplete dominance.

Multiple Choice

4. At one gene locus, there are two alleles, one allele produces black mice, the other produces brown mice. However, there is a second gene locus that when homozygous recessive, suppresses both alleles, producing white mice. This is an example of
   a. pleiotropy.
   b. epistasis.
   c. multiple alleles.
   d. incomplete dominance.
   e. codominance.

5. A single gene affects more than one character. Such a gene is referred to as
   a. pleiotropic.
   b. showing multiple alleles.
   c. codominant.
   d. showing epistasis.
   e. polygenic.

6. In nematodes, a point mutation in a gene causes a 50% increase in life span. It later is discovered that the same mutation also causes an increased resistance to high temperatures and an increase in resistance to damage by free radicals. The gene containing the point mutation must be
   a. polygenic.
   b. codominant.
   c. epistatic.
   d. pleiotropic.

Topic Test 3: Answers

1. **False.** The statement describes multigenic inheritance. Pleiotropy refers to a single gene affecting more than one character.

2. **False.** With incomplete dominance, there is a blending of the phenotypes to produce an intermediate trait. With codominance, the phenotype of the heterozygote shows both traits.

3. **True.**

4. **b.** Here is a case of one gene affecting the expression of another gene, which is epistasis.

5. **a.** By definition.

6. **d.** This is an example of pleiotropy, where one gene is having more than one effect in the organism.

TOPIC 4: PEDIGREE ANALYSIS

KEY POINTS

✓ *What is pedigree analysis?*

✓ *How can pedigree analysis be used to distinguish between dominant and recessive traits?*

Human genetic disorders, and other traits, can be detected and analyzed by examining the pattern of the inheritance of such traits in families. This is called **pedigree analysis**. One use of such analysis is to determine whether an inherited trait is dominant or recessive. This information is important to know when counseling individuals about the odds of their children and grandchildren inheriting the trait.

Figure 12.1 shows typical pedigrees, one for a recessive trait and the other for a dominant trait. You should be able to determine whether a trait is recessive or dominant by analyzing such pedigrees. You also should be able to predict the genotypes of some of the individuals in such pedigree tables.

A dominant trait (allele A) will be expressed when the genotype is either AA or Aa. A recessive trait (allele a) will only be expressed when the genotype of the individual is aa. In Figure 12.1a, we can see that the disorder skipped a generation. Ignoring the possibility of new mutations, only recessive traits can skip generations and then reappear. Each individual who has the trait in Figure 12.1a must have an aa genotype. In the second line, what must be the genotype of the parents who have two offspring with the trait? Since each offspring receives one gene from each parent, each parent must have one copy of the a allele. Therefore, each parent must have the genotype Aa (if they had the genotype aa, the only other possibility, they would have the trait, and they do not).

In Figure 12.1b we can confirm that this appears to be a dominant trait. It does not skip generations—once gone, it does not return in a later generation. Furthermore, it is present in about 50% of the offspring, which suggests that the original father must have had a genotype of Aa. Is it possible that the pedigree shown in Figure 12.1b is for a recessive trait? Yes, it is possible, but unlikely since the father would then have had the genotype aa and the mother the genotype Aa since there are offspring who have the trait (aa). However, then one would have expected three of the four offspring to have the trait, but only two of the four do. The probabilities are not different enough with this small number of individuals to tell for sure. Further data would be necessary, but the pattern is consistent with a dominant trait.

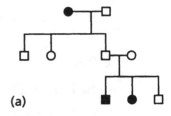

(a)

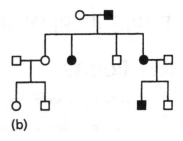

(b)

Figure 12.1 Pedigree analysis. a. Recessive trait. b. Dominant trait. The circles represent females and the squares, males. A filled circle or square indicates the presence of the trait. The vertical lines link parents (above) to their offspring (below).

Topic Test 4: Pedigree Analysis

True/False

1. In Figure 12.1b, assuming that the trait is dominant, individuals with open circles must have a genotype of aa.

2. In Figure 12.1a, the male on the first line probably has a genotype of AA.

3. The three offspring (two males, one female) on the second line in Figure 12.1a must have genotypes of Aa.

Short Answer

4. In Figure 12.1b, given that the trait is dominant, what evidence indicates that the father in the first line is Aa and not AA?

5. In Figure 12.1a, what is the genotype of the daughter, in the middle, on the second line?

6. If the female on the far left of the last line in Figure 12.1b were to ask about the odds of her children having the disorder, what could you say?

Topic Test 4: Answers

1. **True.** Otherwise they would have the trait.

2. **True.** We cannot be sure, but since the female is aa, if the male were Aa, one would have expected one or more of their children to exhibit the disorder.

3. **True.** Since the mother is aa, and the offspring do not have the disorder, they must have inherited an a allele from the mother and an A allele from the father.

4. If the father had been AA, all of his offspring would have had the trait.

5. She must be Aa, since she got one a from her mother, who was aa. She cannot be aa, or she would have the trait, rather than just be a carrier.

6. If the trait is dominant, then she can be confident that her offspring will not have the disorder.

APPLICATION: INTELLIGENCE, A POLYGENIC CHARACTER

We mentioned earlier that intelligence is an example of a polygenic (multigenic) character. Intelligence is influenced by a number of different genes and also by the complex interplay between genes and environment during the development of individuals. Intelligence in an individual can be strongly influenced by such prenatal events as the mother's nutrition and alcohol consumption, and by the early experiences of the infant and child.

We have come to recognize that intelligence itself is a complex character, or actually a set of characters, only a part of which individual tests can measure. Even "intelligence" (IQ) tests are recognized by most to be grossly oversimplified measures. Analyses of the results of IQ tests on populations have allowed us to measure what is called the "heritability" of intelligence. These analyses found that about 80% of the differences in IQ scores among members of typical populations of humans are due to genetic differences.

Some have read this result as indicating that intelligence largely is "fixed" by one's genes, but that would be a mistaken interpretation. The measurement of heritability only holds for the range of environments the population is experiencing at the time of the test. It cannot take into account the possibility of changing intelligence in individuals due to new environments. To the extent that environments change, so can intelligence, and differently for different individuals.

One easy way to see this effect of changing environments is to look at how intelligence scores have changed during this century. Despite a high heritability of intelligence during this century, intelligence test scores have increased dramatically. This is called the Flynn effect, named after the individual who first noted it. Relative to the average score of 100 in 1989, the average individual in 1918 would only have scored 76 on a similarly scaled IQ test. We do not even know for sure what has caused the dramatic rise in IQ scores during this century. Certainly the difference is not genetic, because there have not been enough generations for that kind of difference to arise genetically. So, IQ scores can increase dramatically due to environmental change despite a high, 80% heritability. The interplay of genes and environment can be complex.

Chapter Test

True/False

1. A homozygous individual has the same alleles at the gene locus.

2. A heterozygous individual has two different alleles at the gene locus.

3. There are exceptions to Mendel's rule of independent assortment due to linkage of genes on chromosomes.

Multiple Choice

4. Two plants are heterozygous for flower color. In a cross between the two plants, what will be the fraction of homozygous recessive individuals in the F1 generation?
   a. 0%
   b. 25%
   c. 50%
   d. 75%
   e. 100%

5. Two plants are heterozygous for flower color. In a cross between the two, what fraction will exhibit the recessive phenotype?
   a. 0%
   b. 25%
   c. 50%

d. 75%

e. 100%

6. Green is dominant over clear for the color of the roots of a plant. One such plant whose genotype is homozygous for clear roots is crossed with a plant that is heterozygous. The predicted ratio of root color (green to clear) in the F1 generation is

a. 3:1.

b. 2:1.

c. 1:1.

d. 1:2.

e. 1:3.

7. In incomplete dominance,

a. an intermediate phenotype, between the phenotypes of the two alleles, is observed.

b. the expression of one protein blocks the expression of another.

c. you get more than one effect from each gene, as in sickle-cell anemia.

d. the character is the result of three or more genes acting together.

e. independent assortment does not occur.

8. Polygenic (multigenic) inheritance refers to

a. a single gene affecting more than one character.

b. the case where there are more than two alleles at a gene locus in a population.

c. characters depending on more than one gene.

d. none of the above.

Short Answer

9. Consider the following cross: AABB × AaBb, where A = red fur, is dominant over a = brown fur, and B = black eyes, is dominant over b = green eyes.

a. What is the expected percentage of progeny that are homozygous for red fur color?

b. What is the expected percentage of progeny that are heterozygous for red fur color?

c. What is the expected percentage of progeny homozygous for brown fur color?

d. What is the percentage of progeny that have black eyes?

Chapter Test Answers

1. **T**   2. **T**   3. **T**   4. **b**   5. **b**   6. **c**   7. **a**   8. **c**

9. **a** = 50%; **b** = 50%; **c** = 0%; **d** = 100%

Check Your Performance:

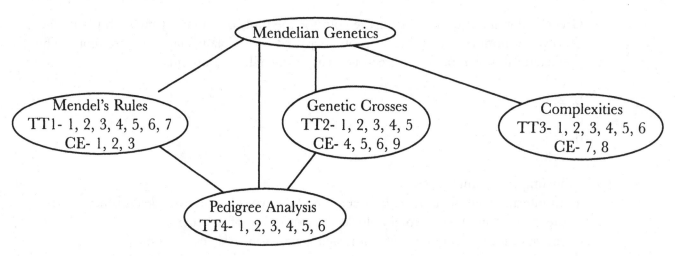

Key: TT = Topic Test; CE = Chapter Exam. Numbers indicate exam questions. Some questions are listed more than once if they refer to more than one topic.

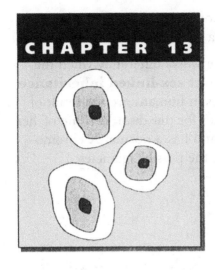

CHAPTER 13

Recombination, Linkage, and Mapping

In the last chapter we examined Mendel's rules of segregation and independent assortment of genes. The segregation and independent assortment of chromosomes occur during meiosis. Thus, at least at the level of chromosomes, Mendel's rules follow from chromosome behavior during meiosis. In this chapter we clarify a few more details about classic genetics. We need to consider sex chromosomes, since inheritance differs somewhat for genes on these chromosomes. We also need to consider the fact that genes can be linked on the same chromosome. We will learn that this leads to exceptions to the idea of independent assortment, but also presents a method for mapping the location of genes on chromosomes.

ESSENTIAL BACKGROUND

- **Sex chromosomes, X and Y (Chapters 11 and 12)**
- **Genetic crosses (Chapter 12)**
- **Crossing-over during meiosis (Chapter 11)**

TOPIC 1: SEX-LINKED INHERITANCE

KEY POINTS

✓ *What is the genetic basis of sex differences in humans?*

✓ *Is it the same in other organisms?*

✓ *What is special about sex-linked inheritance?*

✓ *Can you predict sex-linked, recessive patterns of inheritance?*

As was discussed in earlier chapters, while human females inherit two copies of an X chromosome, males have one X and one Y as their **sex chromosomes**. The X and Y chromosomes have no known genes in common. As sperm are generated during meiosis, each will contain either an X or a Y (the X and Y behave as do other pairs of chromosomes during meiosis, and split from one another during meiosis I). Each ovum contains one of the two copies of the X chromosome from the female. A male's X chromosome always comes from his mother. Whether one is a male or a female depends on the sperm, and whether it contains an X or Y chromosome. Consider some of the old Kings of England (and elsewhere). They wanted male heirs, and sometimes would dispose of wives who gave them females. Now we realize who really should have been disposed of!

Although human females have two copies of the X chromosome, only one appears to be active in most cells. The other X chromosome is condensed, and forms a dark, **Barr body**.

Because males have only one X chromosome, recessive inheritance works differently when the recessive trait is on the X chromosome. You should be able to predict **sex-linked inheritance** patterns. An example will help. Forms of red-green color blindness in humans are the result of alleles on the X chromosome. Consider the case of a female carrier for this disorder. One of her two X chromosomes will have the mutant allele (X*). If her husband has a normal X chromosome, then we can predict the children's genotypes, as usual, from the possible gametes:

	Father	
Mother	**X**	**Y**
X*	X*X	X*Y
X	XX	XY

Among the offspring, half of the females (first column of offspring) will be carriers (X*X), while half of the males (second column of offspring) will have the disorder (X*Y). Even though the gene is a recessive one, the males inheriting one copy will have the disorder because it is the only copy that they inherit. Other sex-linked inheritance problems can be worked in a similar way, as shown in the test questions.

Because males possess only one X chromosome, it seems that they get into trouble more often than females, at least with respect to inheritance of sex-linked disorders. In the case of red-green color blindness, about 6% of males have the disorder, while less than 1% of females do. Red-green color blindness can be a dangerous problem—consider red-green traffic lights. Hemophilia, a condition that causes inadequate blood clotting, is another example of a disorder that is sex-linked recessive, and more common in males.

Other mammals have the same sex inheritance pattern (XX = female, XY = male) that humans do, but this cannot be extended to all organisms. In some birds the role is reversed from that of mammals, since males are ZZ, while females are the ones with different sex chromosomes, ZW. In some insects and nematodes, males are XO (they have only one, unpaired sex chromosome), while females are XX. In ants and bees, males are haploid for all chromosomes, while females are diploid.

Not all species have sex the way that humans do either. Some species are **hermaphrodites**, wherein a single individual makes both sperm and ova. Some nematodes are self-fertilizing hermaphrodites. Sea slugs are quite a different type of hermaphrodite, since during mating they can be male or female, or even act as both at the same time during mating. They can organize into a ring, where each member is both male to the slug on one side and female to the slug on the other!

Topic Test 1: Sex-linked Inheritance

True/False

1. All male animals have XY sex chromosomes.

2. Genetic determination of sex in humans depends on the sperm.

3. Male humans can show a recessive disorder with only one defective copy of an X chromosome gene.

Multiple Choice

4. A color-blind male (sex-linked, recessive) marries a female whose genotype is homozygous for normal color vision. Among their children,
   a. half of the males will be color-blind.
   b. all of the males will be color-blind.
   c. half of the females will be color-blind.
   d. half of the females will be carriers of, but not have, color blindness.
   e. all of the females will be carriers of, but not have, color blindness.

5. A female who is a carrier for color blindness (sex-linked recessive) marries a color-blind male. Among their children,
   a. half will be color-blind, both male and female.
   b. all males and half of the females will be color-blind.
   c. all females and half of the males will be color-blind.
   d. half of all males will be carriers, but will not have color blindness.
   e. none of the males or females will be color-blind.

Topic Test 1: Answers

1. **False.** There are other systems, such as XO, as described earlier.

2. **True.** Presuming that the ovum contributes an X, if the sperm contains an X, the offspring will be XX, or female; if the sperm contains a Y, the offspring will be XY, male.

3. **True.** Males have only one copy of genes on the X chromosome, so even though the allele may be recessive, it is the only one present, and so the only one expressed.

4. **e.** It always helps to set up an analysis, like that given in the text. In this case the only gamete from the female is X, and males will have either X* or Y in the sperm. Thus, all of the males will be normal and all of the females will be carriers, having inherited the X* from their color-blind father.

5. **a.** Again, it is best to structure a table:

	X*	Y
X*	X*X*	X*Y
X	XX*	XY

Half of the offspring, both males and females, will contain copies of only the recessive allele, X*, and so will be color-blind.

TOPIC 2: LINKAGE AND RECOMBINATION

KEY POINTS

✓ *How are genes organized on chromosomes?*

✓ *What are the genetic consequences of the linkage of genes?*

✓ *What is recombination frequency?*

We know that each chromosome contains one long DNA molecule (Chapter 11), and that the sequence of nucleotides along the DNA contains genetic information (Chapter 10). As a consequence we already know about how genes must be organized on chromosomes: They must be **linked** together, like beads on a string, in a linear sequence on each chromosome. A typical chromosome contains hundreds or thousands of genes. In this section we learn that information about the sequence of genes on chromosomes can be obtained because of the existence of crossing-over, the process of exchange of genetic material that occurs in tetrads during meiosis (Chapter 11).

First it is helpful to know the special terminology. This terminology was first developed by T. H. Morgan who was studying mutants in the fruit fly, *Drosophila melanogaster*. He collected fruit flies from the wild, and these were called **wild-type**. Any mutant phenotype was designated by a letter or group of letters. For instance, the wild-type has a gray body, and a recessive mutant has a black body. That mutant would be given the symbol b, lowercase since it is recessive, and the wild-type would be indicated by b^+.

Another mutation in the fruit fly is in wing size. A recessive mutant, vg, has short, vestigial, wings; the wild-type wing is designated by vg^+.

Genes that are linked close enough to each other on the same chromosome do not obey Mendel's rule of independent assortment. Since they are linked to one another, they do not sort independently of one another during meiosis I, as genes on different chromosomes would. Instead, they tend to go to the same daughter cell. In fact, but for crossing-over, they always would. Crossing-over will separate the two genes, if the crossing-over event occurs between them on the chromosome.

Let us compare the expectations with unlinked and linked genes using the two fruit fly mutants, b and vg, just described. Consider the following cross: $(b^+ b \; vg^+ vg) \times (b \; b \; vg \; vg)$. This is a cross between a dihybrid (heterozygote at both gene loci) and an individual that is homozygous recessive at both gene loci. The first individual would be gray and have long wings; the second would be black and have vestigial wings. These are called the **parental types**—the body types (phenotypes) of the two parents. If these two genes were on different chromosomes, we can predict the ratio of different kinds of offspring by establishing a table, based on the possible gametes that each parent can form:

	b vg
$b^+ \; vg^+$	$b^+ b \; vg^+ vg$
$b^+ \; vg$	$b^+ b \; vg \; vg$
$b \; vg^+$	$b \; b \; vg^+ vg$
$b \; vg$	$b \; b \; vg \; vg$

The four different kinds of offspring should be equally abundant, given independent assortment. Two of the four, the first and last rows in the column, have a phenotype like the parents. The other two, in the middle rows, are called **recombinant types**, since they involve a mixing of the two phenotypic traits of the parents. One $(b^+ b \; vg \; vg)$ will be gray with vestigial wings, while the other $(b \; b \; vg^+ vg)$ will have a black body with longer, wild-type wings.

What was actually observed in this cross in the fruit fly was a higher abundance of parental types than of recombinant types. There were only 391 of the recombinant types with 1,909 of the parental types. The reason why there were fewer recombinant types was because the two genes containing the mutations were located near one another on the same chromosome. We say that the genes were linked. Because of the **linkage**, the number of recombinant types were fewer.

The frequency of recombination can be used as a measure of the distance between the two genes on the chromosome. The **recombination frequency** is defined as the frequency of recombinant types divided by the *total* number of offspring, parental plus recombinant types. In our case, recombination frequency = $391/(1,909 + 391) = 391/2,300 = 0.17$ (or 17%). The closer two genes are located on a chromosome, the smaller the recombination frequency will be. The probability of a crossing-over event between the two genes decreases as the separation between the two genes decreases. Genes that are very close can have a recombination frequency close to zero. Notice that because of the way it is defined, the highest that the recombination frequency can be is about 0.50, or 50%. Unlinked genes (on different chromosomes) will have a recombination frequency of 50%. Genes that are on the same chromosome, but are far apart on the chromosome, also can have a recombination frequency of 0.50. This is because of the frequency of crossing-over, which can be high enough to make such genes appear to be unlinked because they will sort independently of one another.

Topic Test 2: Linkage and Recombination

True/False

1. In fruit flies an allele designated as A^+ would be expected to be wild-type, but recessive to the mutant A allele.

2. Recombination frequency is the ratio among offspring of recombinant types divided by parental types.

3. Without linkage, the frequency of the four types of offspring in the table on the fruit flies should be the same. This is because the gametes producing the four types would be equally probable.

Multiple Choice

4. When two genes are linked on the same chromosome, the existence of recombinant types is a result of
   a. Mendel's rule of independent assortment.
   b. crossing-over during meiosis.
   c. sister chromatid separation during mitosis.
   d. a monohybrid cross.

5. Two genes are linked on the same chromosome. One gene specifies hair color and the other specifies eye color. An individual heterozygous at both gene loci is crossed with an individual that is homozygous recessive at both loci. Among the offspring there are 500 individuals of the parental types and 250 recombinants. What is the recombination frequency between the two genes?
   a. 0.50
   b. 0.33
   c. 1.00
   d. 0.67
   e. 0.25

6. Two genes are found on different chromosomes. What will be the recombination frequency between these two genes?
   a. 0.00
   b. 0.33
   c. 0.50
   d. 0.67
   e. 1.00

Short Answer

7. If gene A has a recombination frequency of 0.37 with gene B, and 0.03 with gene C, is gene A closer to gene B or gene C?

Topic Test 2: Answers

1. **True.** The A allele, since it is a capital letter, would be dominant. The "+" indicates that the allele is wild-type.

2. **False.** Recombination frequency is the ratio of recombinant types to the sum of parental plus recombinant types.

3. **True.** Without linkage, the gametes listed in the column would each have a probability of one in four of forming.

4. **b.** Without crossing-over, the genes would remain linked on the same chromosome and the recombination frequency would be zero.

5. **b.** 250/(250 + 500) = 0.33

6. **c.** The expected recombination frequency would be 50% because the number of recombinants = number of parental-type individuals.

7. Gene A would be closer to gene C, since the recombination frequency is smaller. That means there are fewer recombinants between A and C, which would be expected if there is less room between the genes on the chromosome for crossing-over to occur.

TOPIC 3: GENETIC MAPPING

KEY POINTS

✓ *How can one use recombination frequencies to map the location of genes on a chromosome?*

Recombination frequencies give us a measure of how close pairs of genes are on chromosomes. With a little extension, we also have a means of **mapping** sets of genes that are located on a given chromosome. To learn how it works, consider another gene linked to body color in the fruit fly. A mutation in the gene is called cn, and results in brighter red eyes than the wild-type, cn+. The recombination frequency between cn and b is 0.09 (9%). This was determined by the method described in the previous section. Given that the cn-to-b distance is 9% and the b-to-vg distance is 17%, what is the sequence of the three genes on the chromosome? There are two possibilities:

(1) cn---b-------vg   or   (2) b----cn----vg

We know that cn is closer to b, but we do not know what side cn is on, either the same side as vg or the other side. An easy way to determine this is to measure the recombination frequency between cn and vg. In case (1), the answer would be close to 9% + 17%, or 26%. For case (2), the answer would be about 17% − 9%, or 8%. When the cross is done, the answer is 9.5%. Since this is much closer to 8% than to 26%, we can conclude that case (2) is correct.

The reason why the recombination frequency between cn and vg is 9.5% rather than the expected 8% is because of multiple crossing-over events. When estimating recombination frequencies, the longer distances are more likely to be in error because of the possibility of more than one crossing-over event. If two such crossing-over events occur between two genes, they will again become linked, reducing the recombination frequency. Thus, using larger recombination frequencies to estimate map distances on chromosomes will give underestimates of such distances. The figures obtained from crosses covering shorter distances are more reliable.

So, we have a map of these three markers on the *Drosophila* chromosome:

$$\longleftarrow 9 \longrightarrow \longleftarrow 8.5 \longrightarrow$$
$$\text{b---------cn--------vg}$$

Map units, shown above, are simply recombination frequencies expressed as a percentage [one map unit = recombination frequency of 0.01 (1%)].

The same procedure can be used to map other markers, and so to locate other genes, on the chromosome. We then can put together such information to form a map of many of the genes on a chromosome. Other experiments show us that such map distances are not exactly the same as physical distances along the DNA in the chromosome, but the sequencing of the genes is correct, and gives us an approximation of the physical locations of the genes. The reason why the map distances and physical distances do not exactly correspond is because crossing-over is more common in some regions of a chromosome.

Topic Test 3: Genetic Mapping

True/False

1. Each map unit equals 1% recombination frequency.

2. Crossing-over causes recombination among genes on the same chromosome.

Multiple Choice

3. Three genes, A, B, C, are linked to one another with the following recombination frequencies: A-to-B = 4%; B-to-C = 9%; C-to-A = 12%. Which of the genes are closest on a genetic map of the chromosome?
   a. A and B
   b. B and C
   c. A and C
   d. Cannot tell with the information given.
   e. A, B, and C are equally spaced on the chromosome.

4. Recombination frequencies among the three genes are as follows: A-to-B = 17%; A-to-C = 50%; B-to-C = 41%. The most likely ordering of the three genes on a chromosome is
   a. ABC
   b. BAC
   c. Both a and b are equally likely.
   d. None of the above.

Short Answer

5. Four genes, A, B, C, D, are located on the same chromosome. Recombination frequencies obtained from appropriate crosses between gene pairs are as follows:

Gene Pair	Recombination Frequency
A, B	24%
B, C	32%
C, D	22%
A, C	50%
B, D	12%

What is the sequence of genes on the chromosome?

Topic Test 3: Answers

1. **True.** By definition.

2. **True.** Without it, recombination frequencies would be either zero or 50% in all cases.

3. **a.** A and B are only 4 map units apart, B and C are 9, and A and C are 13 (4 + 9, and not 12 because the sum of A-to-B plus B-to-C gives a more accurate estimate than the direct A-to-C does because of multiple crossing-over events).

4. **a.** In working this kind of problem it will help to draw a line for the chromosome and space the genes out along the chromosome according to the map units. Then you will quickly see what sequence works. The sequence ABC accounts for the numbers. Since B is linked to both A and C, A and C must be on the same chromosome even though the recombination frequency between them is 50%, the same as that for unlinked genes.

5. **ABDC.** A-to-B is 24 map units. B-to-C is a further 32, and C must be on the other side of B from A, since A-to-C is 50 map units (unlinked). D must fall between B and C, since it is 22 map units from C and only 12 map units from B. Notice that the B-to-C minus B-to-D does not exactly give the C-to-D distance because of multiple crossing-over events.

OTHER TOPICS

See the web page for details about **extranuclear inheritance**, **inbreeding depression**, and **hybrid vigor**.

APPLICATION: INBREEDING

Most cultures have prohibitions against marriage or mating between close relatives, for biological reasons. Such mating is called **inbreeding**. Most of us carry at least a few recessive alleles that are harmful if expressed. Recessive alleles are not expressed in heterozygous form, but such alleles are more likely to become homozygous if close relatives mate because close relatives are more likely to share alleles for the same recessive traits.

If two individuals are Aa at a given gene locus, then one-fourth of their offspring will be aa. If the resulting recessive trait is harmful, the offspring will suffer the harm.

Inbreeding seems to have played a role in what has become an interesting footnote to a pre-Nazi dream of propagating a "master race." German colonizers from the late 1800s in Paraguay attempted to establish an Aryan utopia. The dream of spawning a superrace fell apart. Those not chased out by malaria-bearing mosquitos, heat, and problems with farmland married among themselves so as not to dilute their racial stock. Instead they suffered the consequences of inbreeding, and their offspring had a high incidence of mental and physical problems. Their desire to produce a genetically superior stock led to the opposite as a result of inbreeding.

Chapter Test

True/False

1. A mutation in a human Y chromosome will cause a genetic change only in males.

2. A mutation in a human Y chromosome will be inherited from father to all of his sons.

3. Females cannot be carriers for mutations in human Y chromosomes.

Multiple Choice

4. Color blindness is a sex-linked, recessive trait. If a male is color-blind and both his father and mother have normal color vision, from whom did he inherit the gene for color blindness?
   a. Mother
   b. Father
   c. Either mother or father
   d. Both mother and father

5. The recombination frequency between genes A and B is 9%. Between A and C it is 5%, and between B and C it is 5%. Which of the following indicates the correct sequence for the mapping of the three genes on a chromosome?
   a. A---B---C
   b. A---C---B
   c. C---A---B
   d. There is no such sequence because 9% recombination frequency is large enough to indicate that B and A are on different chromosomes.

6. Two genes are linked. A cross (doubly heterozygous crossed with doubly homozygous recessive) gives 600 parental types and 200 recombinant types. What is the recombination frequency between these two genes?
   a. 33%
   b. 25%
   c. 300%
   d. 20%
   e. 60%

7. A mother has color blindness, a sex-linked recessive trait. The father has normal color vision. What are the odds of their daughter being color-blind?
   a. 0%
   b. 25%
   c. 50%
   d. 67%
   e. 100%

Chapter Test Answers

1. **T**   2. **T**   3. **T**   4. **a**   5. **b**   6. **b**   7. **a**

Check Your Performance:

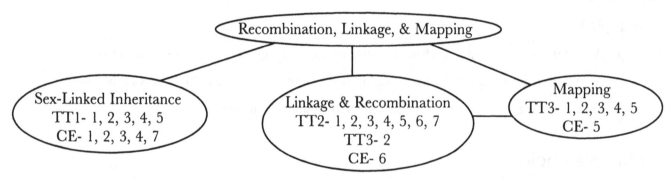

Key: TT = Topic Test; CE = Chapter Exam. Numbers indicate exam questions. Some questions are listed more than once if they refer to more than one topic.

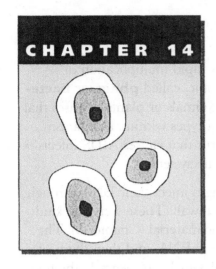

CHAPTER 14

Genetics of Bacteria and Viruses

In this chapter we consider simpler kinds of transfer of genetic information and begin to consider some of the events that control the expression of genetic information. Producing a living organism is not just a matter of having the information for synthesizing proteins. The proteins must be made at the right time, in the right cells.

Bacteria represent life at its simplest. Life still is complicated, but understanding what is happening in bacteria will help us to start to understand what is happening in more complicated organisms. Also, viruses and bacteria have a major impact on eukaryotic organisms, as they cause many diseases. Finally, they are of interest because they have become tools in biotechnology and genetic engineering, as we will learn.

ESSENTIAL BACKGROUND

- DNA (Chapter 9)
- Transcription and translation (Chapter 10)

TOPIC 1: VIRUSES AND THEIR LIFE CYCLES

KEY POINTS

✓ *What do viruses consist of? Are they alive?*

✓ *What is a lytic cycle of virus infection?*

✓ *What is a lysogenic cycle of virus infection?*

✓ *What are restriction enzymes, retroviruses, and reverse transcriptases?*

Viruses cannot metabolize on their own, cannot multiply on their own, and are not composed of cells. They only multiply in the living cells of other organisms. There are different kinds of viruses, which can contain from four to several hundred genes, and which typically are surrounded by a coat of protein. Some **animal viruses** (viruses that infect animals) have an additional coat of membrane around the protein coat. Viruses can be as small as a ribosome or large enough to be seen under a light microscope.

The genes of a virus can be made of DNA or RNA, and can be single stranded or double stranded. Virus coats consist of one, or a small number of, different kinds of proteins. Viruses come in a variety of shapes, some round, some rod shaped, and some with heads (containing DNA) and tails.

The simplest form of virus infection, **lytic infection**, consists of a rapid multiplication of viruses inside a host cell. Most viruses have a narrow host range: Some, called **phage** or **bacteriophage**, infect bacteria. Others are specific for certain kinds of animals or plants. Viruses that infect multicellular organisms typically invade only one or a few cell types within an organism. Cold viruses infect cells in the upper respiratory tract. The HIV virus that causes AIDS infects particular kinds of white blood cells, which form part of our immune system.

The virus must get its genetic material into the host cell. Some phages inject their DNA through virus tails and base plates, which connect the tail to the bacterial cell wall. These were the kind of virus that Hershey and Chase used to study the nature of genetic material (Chapter 9). The injected DNA uses the RNA polymerase of the host cell to make its mRNA, and then the host's ribosomes and other machinery to make the virus proteins, some of which can degrade the host cell's own DNA, while others copy the DNA of the virus, and others serve as coat proteins for new virus.

The viral parts come together either through self-assembly or with the help of a few proteins that put things together. Phage infection takes about a half-hour and generates 100 to 200 viruses from a single virus infecting a bacterium, which bursts open at the end of the infection, releasing the progeny virus. Viruses that only have this lytic type of infection are called **virulent**. In animal cells virus production can last a day or longer, producing and releasing more than a thousand viruses from each cell before it dies.

There are continuing evolutionary battles between phages and the bacteria they infect. Viruses will generate enzymes that selectively degrade the host's DNA, as well as generate protective enzymes that modify the virus DNA by, say, adding extra methyl groups, and avoid self-destruction by the DNA-degrading enzymes. Similarly, bacteria have evolved **restriction enzymes** that destroy DNA by cutting at specific nucleotide sequences not protected by a special modification that the bacterial enzymes will make to the bacterial DNA. Invading virus DNA will be destroyed, but the viruses then evolve such a modification in their own DNA, and on goes the evolutionary battle over time. Humans now make good use of the restriction enzymes that bacteria make, as we will learn in Chapter 15.

Some bacterial viruses, called **temperate phages**, have an alternative life cycle. They undergo a **lysogenic cycle** that reproduces the virus genome without destroying the host. After virus DNA enters the host, it does not immediately start virus replication. Instead, the virus genome, now called a **prophage**, actually is incorporated into the host DNA, remains relatively dormant, and is replicated whenever the host DNA is replicated. In this way the virus spreads as the bacteria multiply. Occasionally, the temperate phages are induced to go through a lytic cycle, producing new viruses and destroying the bacterial cell, as described earlier. Both cycles are available to the temperate phage.

A variety of animal viruses cause disease in humans, ranging from chicken pox, to cold and flu viruses, to herpes and polio. Some animal viruses have RNA as their genetic material and can convert this RNA to DNA by special enzymes called **reverse transcriptases**. These enzymes turn around the usual flow of genetic information, and use RNA as a template for constructing DNA molecules. These viruses are **retroviruses**, and the DNA they make can be integrated into the DNA of the host as a prophage, rather like the temperate phages do in bacteria. Herpes virus and the HIV virus that causes AIDS are examples of retroviruses.

Antibiotics work against bacterial infections, but not against virus infections. Antibiotics typically act on aspects of bacterial metabolism or bacterial structures that differ from those in animals. In contrast to the availability of antibiotics for bacterial infections, at present we cannot do much about some virus infections, other than allow our immune systems to battle the infection (see Chapter 22), but we have been more successful against others. We have made **vaccines** that confer immunity against such viruses as small pox, rabies, and polio. These vaccines have allowed us to eliminate small pox infection throughout the world. Vaccines against polio virus have also allowed us to virtually eliminate polio from many areas of the world. It is more difficult to make vaccines against some viruses because they mutate and evolve too quickly for vaccines to be effective for long.

Topic Test 1: Viruses and Their Life Cycles

True/False

1. Lytic viruses link their genetic material to that of the host.

2. Some lytic viruses destroy the host's DNA.

3. Reverse transcriptase is an enzyme that can make DNA from an RNA template.

Multiple Choice

4. Phages, or bacterial viruses, become integrated into the host cell's DNA during
   a. the lysogenic cycle.
   b. the lytic cycle.
   c. plasmid formation.
   d. intemperate virus infection.
   e. transformation.

5. Viruses
   a. typically contain ribosomes and tRNA.
   b. always contain DNA.
   c. typically consist of nucleic acids only, with no coat, or outer layer.
   d. are only able to multiply in a host cell.
   e. All of the above are correct.

Topic Test 1: Answers

1. **False.** Lysogenic phages do.

2. **True.** Their genetic material codes for enzymes that degrade the host's DNA.

3. **True.** This takes genetic information in a new direction: RNA → DNA.

4. **a.** Such viruses can undergo both lysogenic cycles and lytic cycles.

5. **d.** Viruses do not have the metabolic machinery needed for life, and so "borrow" it from host cells, usually destroying the cell in the process.

TOPIC 2: BACTERIA: GENETIC RECOMBINATION AND PLASMIDS

KEY POINTS

✓ *What three forms of genetic recombination occur in bacteria?*

✓ *What are plasmids?*

The bacteria **Escherichia coli** *(E. coli)* has been widely used by scientists. It thrives in our intestines. Its genetic material, or chromosome, is a single, circular molecule of DNA, about 4 million base-pairs long, containing about 3,000 genes. You can calculate its length (0.34 nanometer/base-pair); then consider that it is folded into a bacterium that is only about 1 micron long. The DNA is over a thousand times as long as the bacterium. Bacterial cells divide by binary fission—splitting in two. First they copy their DNA and sort one copy to each daughter cell, by keeping each DNA linked to the membrane and growing new membrane in between.

You may wonder how evolution occurs in bacteria—what is the source of variation and what are the sources of recombination for bacteria? The variations arise from mutations. There is about one chance in 10 million that a given *E. coli* gene will be mutated in each generation. That does not sound like much until you learn that each of us produces billions of *E. coli* in our intestines every day. Bacterial mutations can make a significant contribution to genetic diversity when such numbers of organisms are being generated.

Three kinds of genetic recombination are known to occur in bacteria. **Transformation** is when bacteria can take up DNA directly from the environment. We saw it at work when we considered the Griffith-Avery experiment (Chapter 9).

Transduction is another form of genetic recombination: Viruses can carry pieces of bacterial genome from one bacterium (where they were generated) to another (which they are infecting). Transduction can be considered an "error" on the part of the virus, as the transduced piece of bacterial DNA often substitutes for a piece of virus DNA; therefore, the virus may not be able to kill its new host because the virus genome is incomplete.

Finally there is **conjugation**, the transfer of genetic material from one bacterium to another through a specialized mating bridge that forms between the two and actually links the cytoplasms of the two cells. Conjugation is facilitated by the existence of special pieces of DNA that bacteria can have in addition to their chromosomes. These pieces, called **plasmids**, usually are small and circular and replicate independently of the chromosome. Some plasmids, called F plasmids, facilitate conjugation. In this way, plasmids can be transferred from one bacterium to another through the mating bridges, which are generated using the genetic information in the plasmid. Bacteria with the F plasmids are called F$^+$.

Plasmids also can become a part of the bacterial chromosome, and as such can also be involved in the transfer of bacterial genes from one bacterium to another through the same mating bridge—the DNA of the chromosome just follows along through the mating bridge after the plasmid it is attached to has passed through. Strains of bacteria with the plasmid as a part of the bacterial chromosome are said to exhibit **high-frequency recombination**, or to be **Hfr** strains.

Topic Test 2: Bacteria: Genetic Recombination and Plasmids

True/False

1. F-plasmids produce Hfr bacteria by becoming a part of the bacterial chromosome.

2. The three forms of genetic recombination in bacteria are transduction, conjugation, and plasmid exchange.

Multiple Choice

3. Conjugation
   a. is the taking up of DNA, free in the environment, by a bacterium.
   b. occurs when a virus takes a piece of DNA of one bacterium into another.
   c. occurs when a plasmid is transferred from one bacterium to another by way of a mating bridge built between the two bacteria.
   d. is the synthesis of DNA on an RNA template.
   e. is the same as translocation.

Short Answer

4. Distinguish between transformation and transduction.

Topic Test 2: Answers

1. **True.** They link to the bacterial chromosome similar to the way that lysogenic phage DNA can.

2. **False.** Plasmids are involved in conjugation. The three are transformation, transduction, and conjugation.

3. **c.** The plasmid is a separate piece of DNA in the bacterium.

4. Both transformation and transduction are forms of genetic recombination in bacteria. Transduction occurs when a bacterium takes up DNA directly from the environment. Transduction occurs when a virus carries a piece of bacterial DNA from one bacterium to another.

TOPIC 3: OPERONS AND THE REGULATION OF PROKARYOTIC GENE EXPRESSION

KEY POINTS

✓ *How is gene expression regulated in bacteria?*

✓ *What are operons, promoters, operators, and regulatory genes?*

Some genes in bacteria are being expressed constantly. That means that mRNA is being made from the DNA by RNA polymerase, and is used on ribosomes to specify protein synthesis

(Chapter 10). Other genes can be turned off or turned on, controlling the amount of gene product that is made in a bacterium. This **gene regulation** enables bacteria to adjust to changes in their environments, such as a change in food source. In bacteria, gene regulation controls the amount of mRNA being made from each gene. This ability to actually govern the level of transcription of genes goes beyond the kind of feedback mechanism that controls the level of activity of individual, allosteric enzymes described in Chapter 6.

Consider the amino acid tryptophan. Most *E. coli* are able to make tryptophan from "scratch," using a set of five enzymes. Tryptophan can feedback-inhibit the first of the five enzymes through allosteric binding.

There is another level of control that governs the actual production of the five enzymes by regulating mRNA production. This can be a more efficient mechanism of control since the cell does not need to expend the energy to make the enzymes if tryptophan is abundant.

The five genes that specify the five enzymes catalyzing tryptophan synthesis are clustered together on the chromosome of *E. coli*. These are called **structural genes**. When tryptophan is needed by the cell, RNA polymerase is able to bind to the promoter region and copy a single mRNA that codes for all five genes. (In prokaryotes, a single mRNA can code for more than one polypeptide.) This is transcription, as described in Chapter 10. The control mechanism involves another gene, called a **regulatory gene**, located elsewhere on the chromosome. The protein that is made by this regulatory gene is called a **repressor**. When adequate tryptophan is present in the cell, some of the tryptophan binds to the repressor, activating the repressor. We say that the tryptophan acts as a **co-repressor**. The activated repressor binds to the **operator** region of the DNA, which is located between the promoter and the structural genes. When the repressor is bound, RNA polymerase cannot make mRNA for the structural genes. The promoter-operator-structural-genes complex on the DNA is called the **trp operon**; trp stands for tryptophan. If the repressor is bound to the operator on the trp operon, we say that the trp operon is "off" and mRNA cannot be produced; otherwise it is "on," and mRNA is made.

Notice how efficient this system is. The enzymes are made if tryptophan is needed, but are not made if tryptophan is not needed. If, say, the human host for the *E. coli* has just had a meal with plenty of protein, and the protein is digested to yield plenty of tryptophan, a bacterium can take up the tryptophan and does not need to expend energy making its own. It will not make its own because the repressor will have tryptophan bound to it and will, in turn, bind to the trp operator, thereby turning the trp operon off—no mRNA will be made so long as the activated repressor is present. The structural gene products, the enzymes, are called **repressible enzymes**, since they are being made in the cell unless their synthesis is repressed by the binding of tryptophan to the otherwise inactive repressor protein. This kind of control mechanism operates for cell products like tryptophan that are needed by the cell.

Another kind of control operates on **inducible enzymes**. Synthesis of these enzymes normally is repressed unless an inducer is present. Such enzymes are found in catabolic (breakdown) pathways, rather than in biosynthetic pathways for molecules such as tryptophan.

Consider the proteins that are required by the bacterium to use lactose, a disaccharide found in milk, as an energy source. If a human host of *E. coli* drinks milk, there will be lots of lactose in the intestine. The bacteria there can break down the lactose, using it both for energy and as a carbon-skeleton source. The group of proteins involved in lactose metabolism have structural genes that are located together on the chromosome, next to their own operator and promoter. This whole team of promoter-operator-structural genes is called the **lac operon**. If there is no lactose around, the protein products are not needed. This is the opposite of the situation with tryptophan.

The regulatory gene for the lac operon makes a repressor protein that binds to the operator region of the DNA, blocking production of mRNA from the lac structural genes. The repressor is active without anything being bound to it. When lactose is present, an isomer of lactose, called allolactose, binds to the repressor and inactivates it—it no longer can bind to the DNA. Then the RNA polymerase can bind to the promoter, pass by the operator region of the DNA, and make mRNA coding for the proteins for lactose metabolism. We call the allolactose an **inducer**. Notice that this, again, is the opposite of the situation for the trp operon, where the repressor only bound when it was activated. Here it binds until an inducer binds to it, which induces the synthesis of mRNA for the lac operon.

Another aspect to the control of the lac operon involves a **catabolite activator protein** **(CAP)**, also known as cyclic AMP receptor protein.

The repressors for both trp and lac operons are allosterically controlled. They bind small molecules that influence their binding to DNA, but in opposite ways.

Topic Test 3: Operons and the Regulation of Prokaryotic Gene Expression

True/False

1. The mRNA for repressible enzymes is not made if a co-repressor is present.

2. The mRNA for inducible enzymes is not made if an inducer is not present.

3. The regulatory gene is a part of the operon.

Multiple Choice

4. Operons in bacteria
   a. include a set of linked genes.
   b. include genes that are regulated together.
   c. include a regulatory site called an operator.
   d. include a promoter region.
   e. include all of the above.

5. Neither tryptophan nor lactose is present in adequate abundance in a bacterium. What is the state of production of gene products for the trp and lac operons?
   a. trp on and lac off
   b. trp off and lac on
   c. trp on and lac on
   d. trp off and lac off

6. A protein produced by a regulatory gene in a bacterium is called
   a. an operator.
   b. a promoter.
   c. an operon.
   d. a repressor.
   e. a co-repressor.

Topic Test 3: Answers

1. **True.** The co-repressor must be bound to the repressor before it will bind to the operator.

2. **True.** The repressor will bind to the operator unless an inducer is present.

3. **False.** The regulatory gene is a part of the control system, but not a part of the operon itself, which consists of promoter, operator, and structural genes.

4. **e.** Together, these are what constitute an operon.

5. **a.** The trp operon is on because tryptophan is needed by the cell. The lac operon is off because these proteins are not needed if no lactate is present.

6. **d.** The regulatory gene specifies the repressor protein. The gene usually is located elsewhere on the chromosome, not at the operon.

APPLICATION: RESISTANCE TO ANTIBIOTICS

Antibiotics help us to overcome bacterial infections. Antibiotics usually act on aspects of bacterial metabolism that differ somewhat from eukaryotic cells, like the differences in ribosomes between prokaryotic and eukaryotic cells that were discussed in Chapter 10. Humans have found in nature, or have developed, chemical antibiotics to help us to overcome bacterial infections.

Unfortunately, in recent years, bacteria have evolved **resistance** to the action of many of these antibiotics. The resistance seems to be spreading rapidly among bacteria. We have learned that special plasmids, called **R factors** or **plasmids**, can carry the genes that confer resistance to antibiotics to the bacteria. A bacterium having an R plasmid can share it with other bacteria through conjugation, as described earlier. The R plasmids can carry genes that confer resistance to several different antibiotics, and so a bacterium can gain resistance to several antibiotics by being the recipient of a single R plasmid.

Antibiotic resistance presents a big problem. We now are finding that many of the antibiotics that we have used for years to combat bacterial infections in humans are not effective for particular patients who have a resistant strain growing inside of them. The resistant bacteria include a number that we thought were under control, but now are presenting us renewed difficulty in treating. Tuberculosis is an example of such a disorder for which resistant strains are appearing.

This resistance might have been foreseen if one were to consider evolutionary processes. Just as phages and bacteria have been "battling" through the evolution of bacterial defense mechanisms, and viral mutations that overcome such mechanisms, so now bacteria are evolving to overcome our use of antibiotics. Notice that we can predict that the more that we use antibiotics, the more rapidly resistant strains will develop. Some now have become concerned about the overprescribing of antibiotics for virus infections in humans, where they do no good. Others express concern about our use of antibiotics in farm animals, which end up in our food chain. Not surprisingly it has been said that a common location for antibiotic-resistant strains is in hospitals, where use of antibiotics is high, and where patients can be more susceptible to infections as well.

Chapter Test

True/False

1. Conjugation is a form of genetic recombination involving plasmids.

2. Conjugation in Hfr strains can transfer bacterial genes from one bacterium to another.

3. Reverse transcriptases are made by retroviruses and use RNA as a template for making DNA.

Multiple Choice

4. Transformation
   a. involves the taking up of bare DNA by a bacterium.
   b. occurs when a virus takes a piece of the DNA of one bacterium into another.
   c. occurs when a plasmid is transferred between bacteria.
   d. is the synthesis of RNA from a DNA template.
   e. is the synthesis of DNA on an RNA template.

5. Concerning gene regulation in bacteria, which of the following is NOT correct?
   a. An operon includes a group of genes with shared regulatory control.
   b. A repressor can bind to the operator region.
   c. A regulatory gene makes the repressor.
   d. All operons are blocked by repressors unless a co-factor binds to the repressor and inactivates it.
   e. Gene regulation occurs primarily at the level of transcription.

6. A lysogenic virus
   a. has an outer coat of membrane.
   b. must use RNA as its genetic material.
   c. can have its genes incorporated into the host's chromosome.
   d. always quickly destroys a cell after infection.

7. The trp operon
   a. is blocked by a repressor if adequate tryptophan is present.
   b. degrades tryptophan.
   c. has no regulatory gene affecting it.
   d. makes one mRNA, which makes one polypeptide chain.
   e. is blocked by a repressor except when adequate tryptophan is present.

Chapter Test Answers

1. **T**   2. **T**   3. **T**   4. **a**   5. **d**   6. **c**   7. **a**

Check Your Performance:

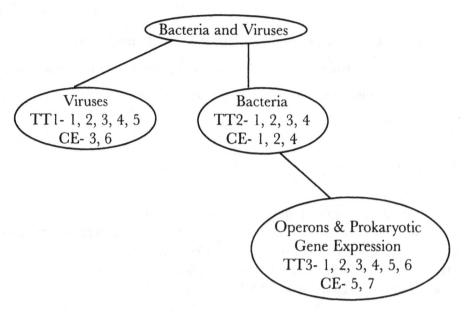

Key: TT = Topic Test; CE = Chapter Exam. Numbers indicate exam questions. Some questions are listed more than once if they refer to more than one topic.

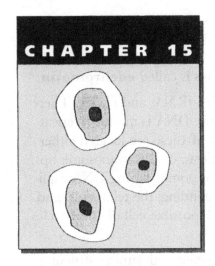

CHAPTER 15

Gene Expression, Gene Technology, and Human Genetics

In the last chapter we considered gene expression in prokaryotic cells. Here, we begin by examining the regulation of gene expression in eukaryotic cells. Eukaryotic cells have added layers of complexity: Their chromosomal organization is more complex, they have nuclei, and their genomes are more complex. Consider a single human cell. It contains 50,000 to 100,000 genes, plus DNA that does not seem to code for anything. In a typical human cell only a few percent of the genes are being expressed at any given time. Different, overlapping sets of genes are being expressed in the different kinds of cells in the body. Embryonic cells may express genes that differ from adult cells. Even a given adult cell will need to turn on or off the synthesis of certain proteins at different times, in response to signals.

In this chapter we also consider human genetics and gene technology. We briefly examine a number of genetic defects in humans. We look at how gene technology might contribute to our understanding and treatment of some of these defects. Gene technology has other uses as well, in fields ranging from agriculture to forensics.

ESSENTIAL BACKGROUND

- DNA (Chapter 9)
- Chromosomes (Chapters 5, 11, and 13)
- Transcription and translation (Chapter 10)
- Restriction enzymes (Chapter 14)

TOPIC 1: REGULATION OF GENE EXPRESSION IN EUKARYOTIC CELLS

KEY POINTS

✓ *What are repetitive sequences and multigene families?*

✓ *What are enhancers and transcription factors?*

✓ *At what levels can gene regulation operate?*

Each chromosome contains one long DNA molecule and many associated proteins, the most common of which are histones. The histones play an important role in the packaging of the DNA within the nucleus. Groups of histones have DNA wrapped around them to form **nucleosomes**. These nucleosomes are like beads on a DNA string, and can be tightly packed together, and then the tight packages can be looped and folded as the chromosome becomes more condensed. In its most condensed state, which occurs during mitosis and in the Barr body (Chapter 13), the chromatin is called **heterochromatin**. In this state the DNA appears to be unavailable for transcription. The unfolded, less compacted form of chromosomes is called **euchromatin**.

In bacterial cells, we learned that most of the DNA codes for mRNA, tRNA, and rRNA. There are other kinds of DNA in eukaryotic cells. About 10% to 20% of the DNA is made of repeating sets of 3 to 10 nucleotides, called **repetitive sequences**. Most of these are located either at the **centromeres**, where they play a role during mitosis and meiosis, or at chromosome tips, called **telomeres**, which play a role in protecting the ends of chromosomes during DNA replication. Special telomerase proteins are involved in duplicating, or extending, the telomeres, and the length of the telomeres can play a role in limiting the number of possible cell division cycles for a cell.

There also are **multigene families** in eukaryotic cells. These are genes with similar structure. For instance, a variety of globin genes code for hemoglobin. Some of the different versions are active in different stages of development. Multigene families probably had a common ancestor gene that duplicated repeatedly during evolution. The duplicated genes then diverged in sequence as mutations occurred, and new functions developed for the duplicated genes during continuing evolution.

The regulation of gene expression in eukaryotic cells differs from that in prokaryotic cells. The RNA polymerase II that makes mRNA in eukaryotic cells requires a few **transcription factors**. These aid in the binding of the RNA polymerase to the promoter regions of the DNA, and some transcription factors bind directly to the promoter region, enhancing the binding of RNA polymerase. Each gene has its own promoter site. A distinctive feature of eukaryotic gene regulation is the presence of **enhancer** sequences on the DNA. These can be a distance away from the actual gene, but the binding of other transcription factors to the enhancer region increases transcription. In Chapter 20 we will examine an example of transcription factors and enhancers. Steroid hormones can bind to receptors that then become transcription factors, binding to enhancer sequences on the DNA and activating expression of particular genes.

Operons have not been found in eukaryotic cells. The genes involved in a given metabolic pathway need not be located together and even can be distributed on different chromosomes. Each gene has its own promoter, but functionally related genes can have common control elements to coordinate their expression.

As with prokaryotic cells, transcription is a major site for the regulation of gene expression in eukaryotic cells, but there are other levels as well. We already learned of the blocking of mRNA production in heterochromatin. In addition, the average life span of mRNA molecules, once synthesized, can be a control point. For example, as mammals begin to lactate (i.e., to produce milk), there is a hormone, prolactin, that causes a 25-fold increase in the lifetime of the mRNA molecules for casein, a major milk protein.

Masking of mRNA also can occur. After synthesis of mRNA, translation can be blocked by masking the mRNA. Unfertilized ova have such masked mRNA, which is unmasked when fertilization occurs.

After protein synthesis, the activity of genes can be regulated by the need for cleavage (splitting), for modification (phosphorylation, etc.), or for co-factors (allosteric enzymes, etc.).

It should be clear that the regulation of gene expression in eukaryotic cells is complex and can occur on a number of different levels.

Topic Test 1: Regulation of Gene Expression in Eukaryotic Cells

True/False

1. Enhancer sequences in DNA can increase transcription rates if they have transcription factors bound to them.

2. Telomeres are short, repeating sets of nucleotides located at the ends of chromosomes.

3. Euchromatin is highly condensed chromatin.

Multiple Choice

4. Gene regulation in eukaryotic cells
   a. can occur at a number of different levels during transcription and translation.
   b. can include the action of transcription factors acting on RNA polymerase, promoters, or enhancers.
   c. can involve enhancer sequences not located right next to genes.
   d. can do all of the above.

Short Answer

5. During evolution, what might be the advantage of having multigene families?

Topic Test 1: Answers

1. **True.** Somehow the binding of protein factors to the enhancer sequences triggers an increase in the rate of transcription.

2. **True.** Such telomeres are made by telomerase.

3. **False.** Heterochromatin is highly condensed chromatin.

4. **d.** Different transcription factors can indeed act in the different sites indicated, and enhancer sequences need not be located right next to a gene. Instead, the DNA probably folds or bends to allow the enhancer sequence with attached transcription factor to bind to, or otherwise stimulate, the RNA polymerase attached to the promoter region.

5. Multigene families allow for the evolution of groups of genes with related functions. The duplication of a gene followed by the ability of the pair to evolve separately allows for one to retain an original function while a second is modified to fill a new role in the cell. Large numbers of such families of related genes are found in eukaryotic organisms.

TOPIC 2: HUMAN GENETIC DISORDERS

KEY POINTS

✓ *What are some of the genetic disorders present in human populations?*

✓ *What causes Down syndrome?*

We previously considered a few genetic disorders in humans. Now we look at a few more. While some genetic disorders cause very direct and obvious phenotypic changes, it is important to realize that most human characteristics are the result of complex interplays between many genes and environmental influences. For example, height is influenced by a number of genes as well as by diet during development. Even in the elderly, height can be influenced by environmental factors such as exercise and calcium levels in the diet, as well as by hormone treatment in post-menopausal women.

Genetic defects need not be in a single gene. For instance, some are the result of chromosomal abnormalities. The most common of these is **Down syndrome**, which results from an extra, third copy of chromosome 21, a **chromosome 21 trisomy**. This is caused by a nondysjunction (nonseparating) of chromosome 21 during meiosis in the female. As a consequence, the resulting ovum ends up with two copies of chromosome 21. During fertilization, a third copy is added from the sperm. Individuals with Down syndrome exhibit, among other things, mental retardation, shortened life span, and heart defects. All of this results from a gene dosage effect—each cell has an extra copy of each of the genes on chromosome 21.

The risk of having an infant with Down syndrome increases with the age of the mother, from less than one chance in a thousand at age 25 to one chance in 32 at age 45. This is one reason why many pregnant women over the age of 35 choose to have **amniocentesis**, a procedure that involves removing amniotic fluid, usually during ultrasound imaging of the fetus, and karyo-typing (checking the number and form of chromosomes). For a further description of the suspected causes of the increase in likelihood of Down syndrome with age of the mother, and of birth defects more generally, see the web page.

There are a variety of recessively inherited traits in humans. Some are deadly; some are innocuous; some are helpful. We focus on a few that generate disorders. One of every 25 white individuals is a carrier for the recessive allele that causes **cystic fibrosis**. It is much rarer in other groups. The allele produces a defective membrane protein that is involved in pumping chloride ions (Cl^-) across the membrane. As a result, in homozygous recessive individuals, chloride accumulates in cells. The higher internal salt concentration causes water to be taken up by osmosis; the mucous coat outside of cells, such as in the lungs, thickens, and this favors the development of pneumonia and other infections. Even with the best treatments available today, individuals with cystic fibrosis usually only live to their late 20s.

Why is the cystic fibrosis allele so common? Why hasn't natural selection weeded it out of white populations? It appears that in heterozygotes, it may reduce the severity of cholera, a deadly disease that was common in white populations until more recently. As natural selection would predict, there appears to be a reason why the gene contributes to fitness in heterozygotic individuals, which offsets the selective disadvantage in homozygotes.

We already discussed another lethal recessive, Tay-Sachs disease (Chapter 5), which is common in Jewish individuals from central Europe, with a frequency of about 1 in every 3,600 births. We also discussed sickle-cell anemia (Chapter 10), present in 1 in every 400 blacks. In Chapter 13 we discussed the sex-linked, recessive disorders of color blindness and hemophilia.

There also are lethal, dominant disorders. Among these is a late-acting trait, **Huntington's disease**, which causes degeneration of the nervous system, but usually is not apparent until ages 35 to 45. We recently located the gene involved in Huntington's disease, and as a result we can test for the disorder in offspring of those who have Huntington's disease (any offspring have a 50–50 chance of inheriting the dominant allele from the affected parent), but we do not yet have a cure. A restriction fragment length polymorphism (RFLP) marker was used to detect the allele. We discuss RFLPs in the next section.

Amyotrophic lateral sclerosis (ALS) is a motor neuron degenerative disorder, also known as Lou Gehrig's disease, named after the baseball player who had it (before Cal Ripkin, Lou Gehrig held the record for playing the most consecutive days without missing a ball game). ALS afflicts Stephen Hawking, the physicist who studies black holes. The ALS gene has been identified as a superoxide dismutase enzyme, whose normal role is to protect the cell from oxygen free radicals (Chapter 7).

Another disorder, **phenylketonuria**, or **PKU**, is caused by a recessive allele that disrupts the ability to break down the amino acid phenylalanine. PKU can be deadly, but can be controlled by diet. Newborns are tested routinely for this disorder.

Topic Test 2: Human Genetic Disorders

True/False

1. Huntington's disease is a recessive disorder.

2. In cystic fibrosis, a sodium channel is not functioning properly.

3. Phenylketonuria is caused by a metabolic defect related to an amino acid.

Multiple Choice

4. The odds of having a child with Down syndrome
   a. increase with the age of the father.
   b. increase as a product of the father's age times the mother's age.
   c. increase with the age of the mother.
   d. Both a and c are correct.
   e. All of the above are correct.

5. One parent has Huntington's disease. What are the odds of a child developing the same disorder?
   a. Zero
   b. One in four
   c. Two in four
   d. Three in four
   e. One

Topic Test 2: Answers

1. **False.** Huntington's disease is dominant. One copy of the gene is all that is necessary to have the disorder.

2. **False.** It is a chloride pump that is not functioning properly.

3. **True.** The disorder blocks the ability of the cells to metabolize the amino acid phenylalanine. As a consequence, without the proper diet, the concentration of phenylalanine builds in the blood, and can damage the brain and other body tissues, leading to death.

4. **c.** Only the mother's age is significant.

5. **c.** We can presume that the parent is Hh, where H = allele for Huntington's disease (since Huntington's is a rare disorder). Then there will be two possible gametes from that parent—H or h, each equally probable. So the odds that the child will develop the disorder (inherit H) is 50%, or two in four.

TOPIC 3: GENE TECHNOLOGY: CLONES, PLASMIDS, PCR, AND RFLPs

KEY POINTS

✓ *How can restriction enzymes be used to make recombinant DNA?*

✓ *What is cloning and how can plasmids be used in cloning?*

✓ *What is PCR and what is it used for?*

✓ *What are RFLPs and how can they be used?*

Recently, scientists have been successful in identifying the genes involved in a variety of genetic disorders, such as Huntington's disease, ALS, and cystic fibrosis, as described already. Scientists are in the middle of sequencing the entire human genome, and already have completed sequencing the genome of several other organisms. Through genetic engineering, large amounts of certain proteins or peptides have been made for the treatment of some human disorders, such as growth hormone for the treatment of stunted growth. The proteins are made by inserting the correct gene into bacteria and letting the bacteria synthesize large amounts of what is normally a rare gene product. This procedure is called **cloning** of the DNA or gene.

Such advances have been made possible because of advances in gene technology, also known as biotechnology or genetic engineering, which refers to the ability to shuffle genes around, copy them, and study them for applied purposes. A few tools have helped with this.

One consists of **restriction enzymes (restriction endonucleases)**, which recognize specific, short sequences of nucleotides, or base-pairs, within DNA and hydrolyze the molecule at any such sites. Hundreds of different restriction enzymes have been found in bacteria, each enzyme recognizing a particular sequence in DNA. The bacteria use them to destroy invading viruses, as described in Chapter 14.

The restriction enzymes cut the DNA on one strand at a slightly different place from the other strand, leaving a few dangling nucleotides, called "sticky ends." For example, one restriction enzyme would cut the DNA with the following sequence (only the relevant nucleotide bases are shown, the rest are indicated by ----):

```
--------------GAATTC----------
--------------CTTAAG----------
```

to make

```
--------------G          AATTC----------
--------------CTTAA          G----------
```

Notice that on the sticky ends, the sequence on one strand is the same as the sequence, read in the opposite direction, along the complementary strand. These sticky ends of single-stranded

DNA can be used to attach the DNA to another piece of DNA that has been cut by the same restriction endonuclease, since they will contain complementary ends. The two pieces of DNA from different sources will link, and then a ligase (Chapter 9) can be used to bond the two together. This is rather like the reverse of the above reaction. Thus, we can form **recombinant DNA**, that is, a single DNA molecule that has come from two different sources.

Another use for restriction enzymes comes from the fact that each one always cuts DNA at the same, precise sequence. We can use a restriction enzyme to cut DNA from two different sources and then compare the cut-up DNA, looking for similarities in the fragments. This has helped in forensics, in paternity suits, and in finding genetic defects, as we discover shortly when we consider RFLPs.

Recombinant DNA can be used to make more copies of the DNA, or even the gene product if the DNA codes for a protein or polypeptide. The gene is added to a bacterial plasmid, and then the plasmid is inserted into bacteria by transformation. Once inside the bacteria, many copies of the DNA can be made as the bacteria grow and divide. The gene is said to be **cloned**.

Another way of making more copies of DNA is through the **polymerase chain reaction (PCR)**. Any desired piece of DNA can be copied many times in a test tube. We speak of the DNA being **amplified**. For PCR, a DNA segment plus short sequences of DNA primer are used with heat-resistant DNA polymerase enzymes. The DNA is heated to split each DNA double helix into two strands, then the mixture is cooled, and the enzymes make a complementary copy of each strand, using existing strands as templates. Two DNA molecules are made for each one started with, and the reaction can be cycled repeatedly to make as many copies as needed. The scientist who developed the procedure won a Nobel Prize in 1993.

One reason why there have been so many recent announcements of the chromosomal location and nature of the genes involved in different genetic disorders has been the use of **restriction fragment length polymorphisms** (**RFLP**s; pronounced "rif-lips"). Restriction endonucleases will cut DNA at particular sequences, as described earlier. If a closely related DNA has a single nucleotide base-pair change at the restriction enzyme's usual site for cutting the DNA, the DNA will remain intact at that point. As a result the DNA fragment will be longer—the piece of DNA that would have been cut into two now remains one. Such length changes can be detected by separating the DNA fragments by gel electrophoresis. Except for identical twins, we each have RFLPs that are unique because of mutations and the unique mix of genes that we receive from our parents. Therefore, we can compare the DNA from two sources to determine if it is similar or not, in terms of the presence or absence of particular lengths of DNA fragments made by the restriction enzyme. Of course, nothing stops us from using more than one such enzyme on parts of the same DNA sample to get a more complete picture.

Some of the uses of gene technology may be obvious by now. If a little blood or semen is left at a crime scene by a criminal, PCR can be used to amplify the amount of DNA. Then, restriction endonucleases can be used to cut the DNA and the fragment lengths obtained from the crime scene can be compared with those from the suspect. Or, to determine paternity, we can compare the DNA fragments between the father and offspring after cutting with restriction enzymes. One also can use RFLP to help locate genetic disorders, as will be described shortly.

Genetic engineering has other potential uses as well. We have been able to modify plants used as food, by adding genes to give them a longer shelf life, more resistance to insects, and so on. Scientists also are attempting to replace defective genes in humans with copies of the normal allele, but getting the genes into individual cells of adults is proving to be challenging. Genetic engineering, as with any new technology, presents potential problems and risks as well.

Topic Test 3: Gene Technology: Clones, Plasmids, PCR, and RFLPs

True/False

1. Recombinant DNA combines DNA from two sources.

2. Plasmids are used in cloning genes.

Multiple Choice

3. Which of the following mixtures would produce RFLPs?
   a. DNA from two different sources compared after cutting with restriction enzymes (endonucleases)
   b. DNA nucleotides plus DNA polymerase
   c. Two kinds of DNA mixed together with DNA polymerase
   d. RNA plus DNA nucleotides plus reverse transcriptase
   e. DNA from a plasmid plus ligase

4. Cloning a gene in bacteria using a plasmid
   a. can be used to generate many copies of the gene.
   b. is the same as PCR.
   c. is the same as RFLP.
   d. cannot yet be done, but we are getting close.

5. A murder has occurred. Samples of blood, some thought to be left by the murderer, but only in small amounts, are being analyzed. Which of the following procedures or tests might be used on the blood by forensic technologists?
   a. PCR
   b. Determination of blood type (A, B, or O)
   c. RFLP analysis
   d. All of the above

Short Answer

*6. Why can't one just put a human gene into a bacterium and obtain the same protein product as would be made by human cells?

Topic Test 3: Answers

1. **True.** DNA from two different sources are combined into a single molecule.

2. **True.** Recombinant DNA in plasmids is introduced into bacteria.

3. **a.** RFLPs are generated by cutting DNA with restriction enzymes.

4. **a.** As the bacteria multiply, so do the copies of the gene in the plasmid.

5. **d.** The PCR increases the amount of DNA in the small samples; the blood type and RFLP analysis help to identify the individual who committed the crime.

6. Because human cells are eukaryotic, their genes contain introns, which need to be spliced out to form the mRNA. Bacteria do not have the necessary machinery, and so would copy the human gene into an mRNA that would contain introns, and thus not make the correct protein. The solution to this problem is to use human mRNA, rather than human genes, as the source of the nucleotide sequence.

APPLICATION: SCREENING FOR GENETIC DISORDERS

RFLPs can be used to locate defective genes on chromosomes and then become a tool in screening for those genetic disorders. Either the altered sequence causing the allele, or a nearby altered gene sequence, can be identified by the particular fragment it produces with a restriction enzyme. For instance, a slight genetic modification can either produce a new restriction enzyme site or eliminate one that was present before the mutation.

If the change is the elimination of a site, what was two different fragments now becomes one (one less cut is made in the DNA), and that new, longer fragment appears at a new place in a gel separating the fragments. This kind of change can occur in DNA and need not be at the exact site of the change that produced the genetic defect. So long as the two are very nearby, they usually will be inherited together (it would require a crossing-over between the two sites during meiosis to separate them).

If the altered sequence is shared among most individuals who have the genetic disorder, then it becomes a marker that can be used to identify the location of the defect and also to identify those individuals likely to have the disorder. This is how the gene for Huntington's disease was located, and a genetic test for the presence of the Huntington's gene was developed. The presence of the marker gives about a 90% certain identification of those who have inherited the Huntington's gene from the parent with the disorder.

Chapter Test

True/False

1. RFLPs are a tool used in screening for genetic disorders.

2. Genes, or gene products, can be produced by cloning.

3. Many copies of a gene can be produced by PCR.

Multiple Choice

4. The chromosomes in a human cell contain about how many genes?
   a. 500
   b. 5,000
   c. 50,000
   d. 500,000
   e. 5 million

5. Which of the following is found in prokaryotes, and not in eukaryotes?
   a. Enhancer genes
   b. Introns
   c. Multigene families
   d. Operons
   e. RNA polymerase

6. Down syndrome
   a. is caused by a chromosome 21 trisomy.
   b. occurs more frequently in offspring of older women.
   c. Both of the above are correct.
   d. None of the above are correct.

7. Concerning restriction enzymes, which of the following is NOT correct? Restriction enzymes
   a. cut DNA at specific sequences of nucleotides.
   b. can be used to make recombinant DNA.
   c. make DNA copies during PCR.
   d. are used to generate RFLPs.

Short Answer

8. A paternity suit claims that George is the father of Sally. RFLP analysis is done. If George is the father, should their RFLP patterns be identical?

Chapter Test Answers

1. **T**   2. **T**   3. **T**   4. **c**   5. **d**   6. **c**   7. **c**

8. **No.** About half of the child's fragments should come from the father, and half from the mother, so there will be similarities and differences. One can ask whether the father accounts for half of the changes, and whether these changes are not accounted for by a sample from the mother.

Check Your Performance:

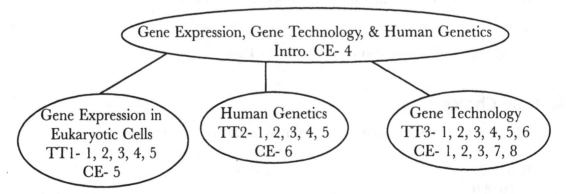

Key: TT = Topic Test; CE = Chapter Exam. Numbers indicate exam questions. Some questions are listed more than once if they refer to more than one topic.

Unit II Exam

Multiple Choice

1. How many different codons are there in the genetic code?
   a. 3
   b. 12
   c. 16
   d. 20
   e. 64

2. During the elongation cycle of translation,
   a. aminoacyl-tRNA enzymes attach amino acids to tRNA.
   b. an uncharged tRNA is released from the ribosome.
   c. translocation occurs.
   d. an incoming "charged" tRNA binds to the ribosome.
   e. b, c, and d are correct.

3. Experiments demonstrating that DNA was the genetic material were completed
   a. during the early 1800s.
   b. during the middle 1800s.
   c. about 1900.
   d. during the 1920s.
   e. during the 1940s.

4. Imagine that a tRNA that normally is linked to the amino acid leucine instead has the amino acid phenylalanine attached. How or where did that error likely occur, and will it be corrected before or during protein synthesis?
   a. The error likely occurred on an aminoacyl-tRNA synthetase and will be corrected when the tRNA binds to the ribosome.
   b. The error likely was the fault of a faulty mRNA molecule and will not be corrected.
   c. The error likely occurred on an aminoacyl-tRNA synthetase and will not be corrected.
   d. The error likely was a result of a faulty rRNA molecule and will be corrected on the ribosome.
   e. This replacement of one amino acid for another is not an error, and so does not need correction.

5. Imagine that life is found on Mars. The genetic material contains three different kinds of nucleotides (bases), and proteins are made from 24 amino acids. The minimum codon length (number of nucleotides in a codon) is
   a. one.
   b. two.
   c. three.
   d. four.
   e. five.

6. Homologous chromosomes
   a. always contain sister chromatids.

b. are defined as the same chromosome, but found in different organisms.

c. are a pair of chromosomes in an individual that contain the same gene loci, but can contain different alleles at each loci.

d. always contain the same alleles.

e. stay together during meiosis I.

7. Comparing meiosis and mitosis, which of the following is correct?

a. Both involve two cell divisions.

b. Both produce daughter cells with a haploid amount of DNA.

c. Both involve the formation of tetrads and crossing-over.

d. Both result in genetically identical daughter cells.

e. All of the above are false.

8. Two parents each show the same trait. Two of their four children also have the trait. What can be said about the trait and the genotype of the two offspring who have the trait? (Standard conventions for dominant and recessive traits are used below.)

a. The trait is dominant, and the offspring could be Aa or AA.

b. The trait is recessive, and the offspring must be aa.

c. The trait is recessive, and the offspring could be aa or Aa.

d. The trait is dominant, and the offspring must be AA.

e. None of the above is correct.

9. A man has a sex-linked recessive trait. He marries a woman who is a carrier (heterozygous) for the trait. What are the odds that their daughters will have the disorder?

a. One in one

b. One in two

c. One in four

d. One in eight

e. Zero

10. A dihybrid cross (a doubly heterozygous individual crossed with a doubly homozygous, recessive individual) is done. There are 989 parental types and 995 recombinant types. Which of the following is correct?

a. There is evidence of linkage because the recombination frequency is about 50%.

b. There is evidence of linkage because the recombination frequency is almost 100%.

c. There is no significant evidence of linkage because the recombination frequency is about 50%.

d. There is no significant evidence of linkage because the recombination frequency is about 100%.

e. There is evidence of linkage because the recombination frequency is about zero.

11. Four genes, A through D, are linked. Given the following recombination frequencies, determine the sequence of the genes on the chromosome:

gene A and gene B: 23%

gene B and gene C: 13%

gene C and gene D: 6%

gene A and gene D: 5%

a. ABCD

b. ACBD

c. ADCB

d. ADBC

e. ABDC

12. If there are 12 chromosomes in an animal cell during the G_1 phase of the cell cycle, the diploid number of chromosomes in this animal is

a. 0.

b. 6.

c. 12.

d. 24.

e. None of the above.

13. Two bacteria form a mating bridge and one sends genetic material to the other through the mating bridge.

a. This statement is false, as such a mating does not occur in bacteria.

b. This is called conjugation, and involves plasmids.

c. This is called transduction, and involves plasmids.

d. This is called transformation, and involves viruses.

e. This is not a form of genetic recombination because the two bacteria always have the same genetic information.

14. The lac operon

a. is blocked by a repressor that is always active.

b. is blocked by a repressor only if a co-repressor is bound to the repressor.

c. is blocked by a repressor unless another molecule is bound to the repressor.

d. has a repressor that is identical to the repressor for the try operon.

e. Both b and d are correct.

15. What are enhancers and how do we think they work?

a. Enhancers are proteins that increase the activity of other proteins by binding at allosteric sites.

b. Enhancers are RNA molecules that speed the process of protein synthesis by binding to ribosomes linked to rough endoplasmic reticulum.

c. They are proteins that speed DNA replication by helping to unwind the DNA.

d. They are cytoskeletal elements that strengthen the shape of the cell by linking to the cell membrane.

e. They are regions of DNA that can bind transcription factors and enhance the rate of transcription for a gene.

16. Concerning human genetics, which of the following is correct?

a. A gene allele that is lethal when homozygous can have positive effects on survival when heterozygous.

b. Down syndrome is the result of a recessive, point mutation.

c. Down syndrome is the result of a dominant, point mutation.

d. The polymerase chain reaction is used to blow up cells.

e. Gene technology is useful for the police, but not for agriculture.

17. A homozygous, recessive individual is crossed with an individual who is heterozygous at the same locus. Among the offspring, what fraction will be homozygous recessive?

a. 0.25

b. 0.50

c. 0.75

d. 1.0

e. None

18. Concerning RFLPs,

    a. they are cut out of DNA by restriction enzymes that come from bacteria.

    b. different restriction enzymes can be used to generate different sets of RFLPs.

    c. they can be used to compare DNA from different sources.

    d. they can aid in the identification of individuals who have particular genetic disorders.

    e. all of the above are correct.

Short Answer

19. Consider the following dihybrid cross with unlinked genes showing dominant/recessive inheritance: aaBB × AaBb, with A = green, a = red, B = big, and b = small.

    a. Indicate the possible genotypes of the offspring from this cross, and indicate their ratio (expected frequency among offspring).

    b. What are the possible phenotypes, and what are their ratios?

20. In the Hershey-Chase "blender" experiment, what radioactive element was used to label the protein in the virus, and why was that particular element used?

21. What are introns?

Unit Exam Answers

1. **e** 2. **e** 3. **e** 4. **c** 5. **c** 6. **c** 7. **e** 8. **a** 9. **b** 10. **c** 11. **c** 12. **c** 13. **b**

14. **c** 15. **e** 16. **a** 17. **b** 18. **e**

19. The four possible genotypes are AaBB, AaBb, aaBB, and aaBb, all in equal abundance: 1:1:1:1 (or 1/4 each). The two phenotypes are green-big and red-big, in equal abundance: 1:1 (or 1/2 each).

20. Sulfur was used because it is found in two amino acids used in making proteins but is not an element that is part of the DNA molecule. This helped Hershey and Chase to distinguish the protein from the DNA.

21. Introns are found in eukaryotic organisms, not in prokaryotic ones. They are segments of pre-mRNA molecules that are spliced out as processing in the nucleus of the cell produces mRNA from the pre-mRNA.

UNIT IIII:
EVOLUTION

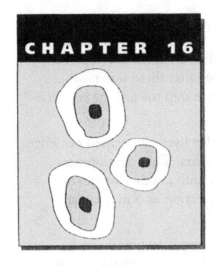

CHAPTER 16

Evolution and Natural Selection

Charles Darwin was not the first to propose that evolution occurs, but he was the first to support it with so much evidence that the vast majority of biologists came to believe it. **Evolution** means that populations of organisms change through generations as a result of changes in the frequencies of inheritable traits in the population. The various organisms around us were not created separately, nor all at once. Instead, they have evolved from common ancestors, with species coming and going, and they continue to do so.

At the time of Darwin, scientists and nonscientists alike strongly held contrary views about the history of life. Such views were rooted deeply in Western culture. They included the ideas that the earth was only a few thousand years old and that it was populated with life-forms (species) that were unchanging. Consider Thomas Jefferson, the former president of the United States, who, in the late 1700s and early 1800s, did research on American life-forms. Jefferson came to know of large, elephant-like fossils called mastodons that had been found in the United States. Jefferson was convinced that mastodons must still be alive in the United States, and expected Lewis and Clark, on their expedition, to find some. But even in Jefferson's time, and especially by the mid-1800s, as Darwin was writing *On the Origin of Species*, the traditional views were not adequate to explain growing evidence to the contrary.

Darwin also proposed a mechanism, natural selection, that underlies evolution. We examine evolution and natural selection in this chapter and the next.

ESSENTIAL BACKGROUND

- Basis of heredity (Chapters 9 and 10)

TOPIC 1: EVOLUTION: HISTORY AND EVIDENCE

KEY POINTS

✓ *What is the history of the idea of evolution?*

✓ *What evidence supports the theory of evolution?*

History

Georges Cuvier, a contemporary of Jefferson, collected fossils and is one of the originators of the field of paleontology, the study of fossils. He saw that new life-forms appeared, and others disappeared, at different layers, or strata, on the earth's surface. He accepted that some species went extinct, but still believed that no new species were appearing. Instead he presumed that the new ones must have migrated from other areas of earth. Since his time we have developed a more complete view of the fossil record around the globe. We can see that there was nowhere for many of the new life-forms to migrate from. Cuvier had taken one step toward the modern view, but strongly opposed the idea of evolution.

Leading geologists of the same era, such as James Hutton and Charles Lyell, developed evidence indicating that the earth might be much older than a few thousand years. That development gave Darwin some support, but he still resisted publishing his results until another researcher, Alfred Wallace, independently developed the same idea of natural selection as a mechanism underlying evolution.

Evolutionary hypotheses predated Darwin, but were not popular. Lamarck proposed one hypothesis, and developed a mechanism that involved the inheritance of acquired characteristics. According to that hypothesis, if one exercised, one's children would be born with bigger muscles. Lamarck's mechanism was faulty, but the idea of evolution was on target. Lamarck came to fame a second time in the Soviet Union in the mid-1900s. Politicians there told scientists that Lamarckian ideas on the inheritance of acquired characteristics were compatible with Communism, and so a whole generation of geneticists wasted their time pursuing a dead end.

In terms of the theory of evolution, we credit Darwin with presenting enough evidence to convince most scientists that it had occurred. Since Darwin's time, the evidence favoring evolution has only grown.

Evidence for Evolution

Evidence from a variety of sources supports both the theory of evolution and the idea that evolution has been underway for billions of years, not thousands. Sometimes the changes that cause extinctions are dramatic and sudden events, such as the meteor suspected to have killed the dinosaurs 65 million years ago, or the more recent massive destruction of ecosystems by humans, which is dramatic and sudden on an evolutionary time scale. Sometimes evolution is more gradual.

One convincing line of evidence favoring evolution comes from **biogeography**, the distribution of species around the world. That distribution is what caught Darwin's attention on the Galapagos Islands. Geographic isolation contributes to the evolution of new species because isolated populations can evolve in different directions. The population on one island could have a different environment from that on another island or on the mainland. As natural selection takes place, the populations will diverge.

There also is the **fossil record** showing the range of organisms that lived in the past, but no longer do, and the rise of new species, families, orders, and classes. The fossil record also suggests linkages among the species, with some forms likely evolving from earlier ones.

Taxonomy offers supporting evidence, allowing one to trace genealogy. Taxonomy is the theory and practice of classifying organisms. The taxonomic tree of life predicts the flow of genes, with branch points on the tree indicating divergence from common ancestors for lines of organisms.

Comparative anatomy reveals similar anatomical structures between closely related species. For instance, in the mammals, the forelimbs of humans, cats, bats, and whales share similarities despite the great differences in usage. These are called **homologous structures**. There also are **vestigial organs**, which are rudimentary structures of little or no use to an organism, but appear to be historical remnants of structures that had a use in ancestors. For instance, some snakes retain remnants of pelvis and leg bones that were useful to walking ancestors. Some whale fossils similarly show small hind legs. The human appendix may also be a vestigial organ.

Study of **comparative embryology**, following the embryonic development of organisms, also reveals the remnants of evolution. Mammals and birds go through an embryonic stage where gill pouches are present. These develop into gills in fish, but not in birds and mammals.

More recently, data from **molecular biology** have supported the relatedness of different species suggested by taxonomy by revealing similarities in the sequence of bases in the DNA, or the sequence of amino acids in the proteins, of different organisms. As expected, more closely related organisms (less time to a common ancestor) share more similarities in nucleotide sequences in DNA and in amino acid sequences in proteins. The relationships between organisms, suggested by taxonomy on the basis of gross physical or behavioral similarities, now have been confirmed and refined by the data of molecular biology.

Evolution has become a well-established scientific theory.

Topic Test 1: Evolution: History and Evidence

True/False

1. Lamarck proposed that evolution had occurred before Darwin did.

2. Biogeography is the study of how species are distributed around the world.

3. Humans evolved from chimps.

Multiple Choice

4. Which of the following gives the strongest evidence in favor of evolution?
    a. There are variations in the amino acid sequence of a protein from different organisms.
    b. There are variations in the sequence of nucleotides in the genes that code for a protein.
    c. The variations in the amino acid sequence of a protein are greater between plants and humans than between mice and humans.
    d. There are variations among humans that are not inheritable.

5. Evidence for evolution includes
    a. the number of differences among species in the amino acid sequences of proteins.
    b. the geographical distribution of species.
    c. the fossil record.
    d. the existence of homologous structures.
    e. all of the above.

Short Answer

6. Has evolution progressed from bacteria at one end to humans at the other?

Topic Test 1: Answers

1. **True.** He did, but he did not have the necessary supporting evidence to convince most others of his views, and his hypothesis concerning how it occurs was in error.

2. **True.** For instance, the isolation of some areas, such as islands, allows for evolutionary divergence in separate populations to occur. Over time, such divergence can lead to the development of new species.

3. **False.** Humans and chimps had a common ancestor about 6 million years ago, but neither species was around then.

4. **c.** The number of differences in the amino acid sequence between two different species increases as the length of time back to a common ancestor increases.

5. **e.** There is support for evolution from all of these areas.

6. While it is true that more complex life-forms have appeared as multicellular, eukaryotic organisms evolved, today's bacteria can trace their evolutionary line back just as far as we can, to the origin of life. Evolution can sometimes push for simplicity as well as for complexity—consider the simply organized bacterial genome.

   Evolution is like a many-branched tree. *Today's* bacteria and humans are at the end of current twigs on the tree. There are millions of such tips, one for each species on earth. There appears to be no good reason to call one tip more evolved than another. Evolution has not progressed from bacteria to humans; both have progressed from earlier forms.

TOPIC 2: NATURAL SELECTION

KEY POINTS

✓ *What is the theory of natural selection?*

✓ *How does artificial selection support natural selection?*

Darwin referred to evolution as descent with modification, indicating that the descendants of organisms accumulated modifications, or adaptations, which allow them to "fit" better in their environments. Darwin and Wallace proposed a mechanism that describes how the adaptation to the environment occurs. With **natural selection**, those members of a population that are better adapted to their environment will leave more offspring, and so their heritable characteristics, their genes, will become more common in the population in succeeding generations. Today we know that Darwin's heritable variations are based on different genes, or different alleles of genes. The organisms that are more fit, or better adapted, to their given environment will tend to contribute more to future generations than will the less-fit members of the population. Natural selection presumes the following:

1. Organisms have high (potential) fertility.

2. Natural resources are limited.

3. There are inheritable variations in populations.

We know that organisms have high potential fertility: Exponential growth is possible for organisms, because of reproduction. We also know today that heritable variations are the result of mutations and genetic recombinations. Natural selection concludes that **there will be a selec-**

tion for those heritable variations that allow individuals to contribute more surviving offspring to the next generation.

The potential to produce more organisms than the environment can sustain leads to a struggle for survival, with only a fraction of offspring surviving each generation. Individuals vary significantly in their characteristics. At least some of this variation is heritable, and can contribute to survival. Individuals whose genes best fit their environment will be more likely to survive and leave more offspring than less-fit individuals. This unequal number of offspring will lead to changes in the population, with favorable characteristics, caused by favorable genes and alleles of genes, accumulating over generations.

Natural selection does not require a physical fight between individuals. Usually, it operates in more subtle ways. One individual may be better able to obtain food (a longer neck in a giraffe), avoid predators (better camouflage in a rabbit), attract pollinators (flower color on a plant), or attract a mate (secondary sex characteristics).

Evolution of organisms involves factors other than natural selection, as we will see in the next chapter. However, natural selection is the only mechanism that has a direction to it. As a result of natural selection, organisms become better adapted to their environments. Of course, changes in the environment can bring about changes in what is selected for.

The English peppered moth represents an example of how natural selection works. The moth has light and dark variants, determined by heredity. Before the industrial revolution in Britain, the light-colored moth was common, and the dark variety was relatively rare. The light moth was well camouflaged on light-colored tree trunks. With the industrial revolution, most tree trunks became darker in color, and the population of peppered moths evolved to be predominantly dark colored. We presume that a major reason was the predation by birds of the light-colored moths resting on dark tree bark. Recently, as Britain's pollution is reducing, and tree trunks are becoming lighter, the proportion of light-colored peppered moths is increasing again.

Additional evidence supporting the theory of natural selection has been obtained by experiments in **artificial selection**. Scientists are able to impose artificial selection on a population and study the resulting changes by selecting for particular traits in a population. In fact, Darwin realized that such artificial selection had already been carried out in agriculture, producing better fruits, vegetables, and farm animals. Race horses had been bred to be faster than wild horses. Broccoli, cauliflower, cabbage, kale, and brussels sprouts all have a common ancestor, a wild mustard plant. Humans selectively bred the wild mustard plant to emphasize different parts, evolving these different foods through artificial selection.

We will examine more details about natural selection in the next chapter as we examine microevolution.

Topic Test 2: Natural Selection

True/False

1. The potential for overpopulation contributes to the driving force of natural selection.

2. Limiting resources contribute to the struggle, or competition, among individuals for those resources.

3. There can be variations among individuals that contribute to survival, but are not heritable.

Multiple Choice

4. Which of the following is a premise underlying natural selection?
   a. The inheritance of acquired characteristics
   b. That the earth is about 6,000 years old
   c. The existence of excess resources
   d. The existence of heritable variations in a population
   e. All of the above

Short Answer

5. Considering natural selection, is it true that a complex, organized system requires a designer?

Topic Test 2: Answers

1. **True.** The continuing production of offspring increases the competition for resources. Such resources can include food, shelter, and territory.

2. **True.** As resources become more limited, the intensity of competition for the resources increases.

3. **True.** Not all variations are the result of genes. Environment and experience can contribute to survival.

4. **d.** Heritable variations are necessary if natural selection is to have something to act on.

5. Natural selection allows for a driving force that can produce organisms that are more fit. Such fitness can include the development of organized systems, where such systems contribute to survival of the individual and the number of viable offspring that the individual produces. Thus, it is not necessary to have a designer for complex systems; natural selection can substitute for such a designer.

APPLICATION: INCREASING LIFE SPAN BY ARTIFICIAL SELECTION

A recent contribution to our understanding of why we age comes from experiments in evolution. As we have seen, natural selection maximizes the ability of organisms to have offspring for the next generation. However, the theory of natural selection would suggest that we can select for other, artificial criteria by suitable adjustments in the environment.

Several laboratories have achieved significant increases in life spans for fruit flies by using special breeding conditions. In several of these studies, late offspring were selected, artificially, for further breeding. Thus, the offspring from female flies who had offspring later than typical flies were chosen and allowed to undergo reproduction. This was continued for many generations. By selecting late offspring, the scientists were selecting for heritable traits in the parents that allowed them to continue to have offspring late in life. After several dozen generations undergoing such selection, flies were able to generate offspring at later and later times. In addition, the average life span for the selected animals also increased.

The "old" strains lived up to 50% longer than the unselected animals. The increased longevity in these strains was accompanied by several other changes in the animals. They grew somewhat larger as larva and tended to take longer to reach adulthood. The "old" animals also were more resistant to stresses, such as starvation and heat. It is as if a larger, more durable body was developed at the cost of a later initiation of reproduction. This kind of selection would not occur naturally because the delay in reproduction would place the flies at a selective disadvantage.

Chapter Test

True/False

1. Georges Cuvier studied fossils.

2. Alfred Wallace was the co-discoverer of natural selection.

3. Thomas Jefferson thought that mastodons still roamed the earth.

Multiple Choice

4. Charles Darwin
    a. was the first to propose that evolution was occurring.
    b. proposed his theory even though he did not know about supporting evidence from biogeography.
    c. disagreed with Wallace about the mechanism underlying evolution in living organisms.
    d. was the first to put together a mass of evidence in favor of evolution, while offering a mechanism explaining how evolution occurred.
    e. did all the above.

5. Which of the following is NOT one of the premises underlying natural selection?
    a. The potential for overproduction (exponential growth)
    b. Existence of heritable variations in a population
    c. Presence of limited resources
    d. A physical battle among offspring, where some are injured or die

6. A dolphin's flipper and a bat's wing are examples of
    a. vestigial organs.
    b. homologous structures.
    c. fossils.
    d. support for evolution through comparative embryology.

Short Answer

7. What does natural selection do?

Chapter Test: Answers

1. **T** 2. **T** 3. **T** 4. **d** 5. **d** 6. **b**

7. Natural selection increases the frequency of particular genes or alleles in a population, through generations, because such heritable variations contribute to the fitness of the organism. By fitness we mean the likelihood of success in leaving more surviving offspring relative to other members of the population. The result is populations of organisms that are better adapted to their environments.

Check Your Performance:

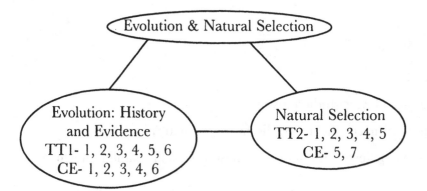

Key: TT = Topic Test; CE = Chapter Exam. Numbers indicate exam questions. Some questions are listed more than once if they refer to more than one topic.

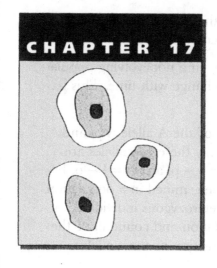

CHAPTER 17

Microevolution and Population Genetics

Populations are the smallest units that can evolve, but the forces that influence populations act on the individuals in the population. For example, individual pepper moths do not change color to match tree bark. Instead, by natural selection, the best-camouflaged ones live longer and produce more offspring. The population evolves, with the more fit individuals being favored.

While natural selection is a major force acting to shape genotypes, other factors also can influence the relative abundance of different alleles in populations. We explore these factors in this chapter.

ESSENTIAL BACKGROUND

- **Natural selection (Chapter 16)**
- **Mutations (Chapter 10)**
- **Heterozygotes (Chapter 12)**
- **Genetic recombination (Chapters 11–13)**

TOPIC 1: MICROEVOLUTION AND HARDY-WEINBERG EQUILIBRIUM

KEY POINTS

✓ *What are populations, gene pools, and species?*

✓ *What is Hardy-Weinberg equilibrium, and what conditions are required for it?*

✓ *What are the five causes of microevolution?*

A **population** is a group of individuals of the same species in a local area. A **species** is a group of populations that have the potential to interbreed in nature. There are more sophisticated definitions of species, but this one will serve for now. The total of all the alleles at all gene loci in all individuals of a population is called the **gene pool** for that population. If all members of a population are homozygous at a given gene locus, that allele is said to be **fixed** in the population. Most species have a considerable number of such fixed alleles. Only a fraction of the gene loci will have two or more alleles present in the population. If the relative frequencies of the different alleles in the gene pool are changing, we say that **microevolution** is occurring.

Macroevolution deals with the evolution of species and beyond, with the way that life has evolved to form orders, classes, phyla, and kingdoms. We focus on microevolution here.

The standard for studying microevolution is a nonevolving population. In a nonevolving population, the frequencies of each of the alleles in the gene pool must not change with time. This is part of **Hardy-Weinberg equilibrium**.

Consider the case of a gene with two alleles, A and a. The frequency of the A allele is p, and the frequency of the a allele is q. We know that $p + q = 1$, because only these two alleles are present, so the sum of the fractions must be 1. We then know that $q = 1 - p$, and $p = 1 - q$, by simple math. You need to be clear about how the allele frequency is determined in a population. A homozygous dominant individual, AA, has two alleles, both A. A heterozygous individual has two alleles, A and a. If we go through each member of the population and count all of the A alleles and a alleles, then the fraction of A alleles gives p, and the fraction of the a alleles gives q.

If a population is in Hardy-Weinberg equilibrium, knowing the value of either p or q allows us to determine the fraction of individuals who are homozygous and heterozygous in the population. For instance, if $p = 0.8$, then $q = 1 - p = 0.2$. Furthermore, because we are at equilibrium, we can view the entire gene pool as randomly mixing each generation. Therefore, we determine the frequency of the AA (homozygous dominant) genotype in the population by calculating the probability of pulling two A alleles out of the gene pool. The probability of pulling one is $p = 0.8$, so the probability of pulling two would be $p^2 = 0.8 \times 0.8 = 0.64$, or 64%.

Similarly, the frequency of the aa (homozygous recessive) genotype is $q^2 = 0.2 \times 0.2 = 0.04$, or 4%. The frequency of heterozygotes is a little more subtle because there are two genotypes for being heterozygote, Aa and aA. That is, we can first pull an A allele and then an a allele out of the gene pool, or we can pull an a allele and then an A allele (in other words, the sperm can be A and the ovum a, or the sperm can be a and the ovum A). Therefore, the frequency of heterozygotes in the population is $2pq = 2 \times 0.8 \times 0.2 = 0.32$, or 32%. We can check the math by adding these three probabilities together: 64% + 4% + 32% = 100%, as expected since these are the only possibilities. It might be helpful to notice that the sum of the terms we used, $p^2 + 2pq + q^2$, equals $(p + q)^2$.

If a population is in Hardy-Weinberg equilibrium, five factors must be true:

1. Very large population

2. Isolation from other populations

3. No net mutation

4. Random mating

5. No natural selection

If all five hold, there will be no change over time in the fraction of alleles present at each gene locus. If one or more of these five factors fail to hold, then microevolution can occur. Now, we consider violations in each of the five factors; these produce violations of Hardy-Weinberg equilibrium.

1. Small population. If a population is small, then chance events can cause the frequency of alleles to change with time. This is **genetic drift**. Just through random chance, organisms with

some alleles will produce fewer offspring. It is like flipping a coin—if we flip a small number of times, the ratio of heads to tails can differ from 1:1. If we flip a large number of times (large population), the fraction gets to be very close to 1:1. During periods of small population, there can be dramatic shifts in the ratio of alleles in the population. As long as the population is small, the allele frequency can continue to change because of chance effects.

Genetic drift can be observed whenever populations are small, but there are two times when this commonly occurs. One is when the population hits a so-called **bottleneck**, when a population is drastically reduced in size. Cheetahs had their populations drastically reduced twice, once about 10,000 years ago owing to an ice age and more recently owing to hunting and habitat destruction by humans. The bottlenecks appear to have resulted in a decline in genetic diversity in surviving cheetahs—they are homozygous for most of their alleles, as will be discussed in the Application section of this chapter.

Another form of genetic drift comes as a result of the **founder effect**. One pregnant female, one seed, or a few individuals sometimes colonize an isolated area—an island, lake, or mountain-top—that is new to the species. Those founders are unlikely to have the same frequencies of alleles as the entire population that they came from, and the size of their population at the new site will be small. Genetic drift can occur, and Hardy-Weinberg equilibrium will not hold. Such drift can continue until the population gets large.

2. A second kind of violation of Hardy-Weinberg equilibrium is caused by migrations between populations. This is called **gene flow**. Migrations, wind, or ocean currents can move individuals from one population to another. Both populations undergo microevolution as a result of the gene flow. Of course, if the gene flow is great enough, it can lead to the two populations becoming combined. Gene flow tends to reduce any differences between populations.

3. A third kind of violation is caused by **mutations**. New mutations appearing in a population produce changes in the gene pool: One allele becomes another. Thus, the population is not in Hardy-Weinberg equilibrium. In large populations, most mutations have little effect because they are rare. But if the mutation is favorable, it can increase rapidly due to natural selection.

4. Mating preferences of certain phenotypes in a population also violate Hardy-Weinberg equilibrium. This **nonrandom mating** is not natural selection. It does not involve one individual mating while another does not. Instead, for example, taller females mate more frequently with taller men than with shorter men. The shorter women mate just as frequently, just with shorter men, on the average. This form of nonrandom mating is called **assortative mating**. What changes as a result of assortative mating is not the frequency of alleles, not p or q, but the proportion of heterozygotes and homozygotes that will not follow Hardy-Weinberg equilibrium. The probability of AA will not be p^2, because an individual who is AA has a higher probability of mating with another AA than chance alone would predict. The allele frequency in the gene pool does not change, but the proportion of heterozygotes will decrease.

Another kind of nonrandom mating is **inbreeding**. Sometimes there are neighborhoods within which members of a population are more likely to mate with each other. Again, this will cause a deviation from the expected fraction of homozygous and heterozygous individuals because organisms in the neighborhood will tend to be more alike in genotype.

5. **Natural selection** is a violation of Hardy-Weinberg equilibrium. Selection is adaptive. It results in an increase in the frequency of certain alleles as a population becomes better adapted to the environment.

Topic Test 1: Microevolution and Hardy-Weinberg Equilibrium

True/False

1. Factors other than natural selection can influence gene pools.

2. Natural selection can occur during Hardy-Weinberg equilibrium.

3. Inbreeding is an example of genetic drift.

Multiple Choice

4. Hardy-Weinberg equilibrium requires which of the following?
   a. Natural selection
   b. Nonrandom mating
   c. Significant migration between populations
   d. Large populations
   e. Genetic drift

5. Which of the following is correct?
   a. Individuals evolve.
   b. Natural selection acts on populations.
   c. The selective forces that influence population gene pools act on individuals.
   d. Natural selection acts on nonheritable phenotypic variation.
   e. Natural selection is not a cause of microevolution.

6. A population that is NOT in Hardy-Weinberg equilibrium contains 25% homozygous dominant individuals. What fraction of individuals in the population are heterozygous?
   a. 25%
   b. 50%
   c. 75%
   d. 12.5%
   e. Cannot tell from the information given.

Short Answer

7. A population is in Hardy-Weinberg equilibrium. At one gene locus, two alleles, A and a, are present. The frequency of the a allele is 0.7.
   a. What fraction of individuals in the population are homozygous dominant at this gene locus?
   b. What fraction are heterozygous?

Topic Test 1: Answers

1. **True.** The factors include genetic drift, mutations, and gene flow.

2. **False.** During natural selection, gene pool allele frequencies change, a violation of Hardy-Weinberg equilibrium.

3. **False.** Inbreeding is an example of nonrandom mating.

4. **d.** The other answers all would violate Hardy-Weinberg equilibrium.

5. **c.** Populations evolve; natural selection acts on individuals.

6. **e.** Since the population is not in Hardy-Weinberg equilibrium, one would need more information to answer the question.

7. **a.** The fraction of individuals that are homozygous dominant is the square of the fraction of homozygous alleles in the gene pool. That fraction, $p = 1 - q = 1 - 0.7 = 0.3$. So the fraction of individuals that are homozygous dominant is 0.09, or 9%. b. The fraction that is heterozygous is $2pq = 2 \times 0.3 \times 0.7 = 0.42$, or 42%.

TOPIC 2: GENETIC VARIATION IN POPULATIONS

KEY POINTS

✓ *How is genetic variation maintained with natural selection?*

✓ *What are heterozygote advantage, balanced polymorphism, and frequency-dependent selection?*

✓ *What is relative fitness?*

✓ *What are stabilizing, directional, and diversifying selection?*

✓ *Why can't natural selection make perfect creatures?*

Natural selection depends on genetic variations. There must be different genotypes among the individuals in a population for natural selection to occur. Just having variations in a population is not enough, since the variations must be heritable for natural selection to work. What maintains such genetic variability in a population? Why hasn't natural selection already eliminated all but the one, very best, genotype for a population? We learn why in this section. The sources of genetic variation are mutation and genetic recombination. Considerable variation is maintained in many populations despite natural selection, as we will learn.

Most of natural selection is operating on polygenic characters, that is, characters that are influenced by more than one gene. Sometimes such characters, such as height, are relatively continuous. Where there are discrete characters, and there are two or more traits for a character in a population, we use the term **morphs**. Morphs are like traits. A population is called **polymorphic** if it has two or more morphs for a character. Polymorphism is common in most populations, including humans. Freckles and blood groups are polymorphic characters in humans. In a population of fruit flies, about one-third of all genetic loci have more than one allele, and in individual fruit flies, typically, one in 10 loci is heterozygotic. Humans probably are similar. The high abundance of heterozygotic characteristics and multiple alleles in most populations provides a range of genotypes and phenotypes, which get remixed at each mating.

Why doesn't natural selection "weed out" genetic variation? Because organisms have features that help to preserve genetic variation. One of the more important is **diploidy**. In diploid organisms, recessive alleles can persist in heterozygotic form.

Natural selection actually can preserve variation under certain conditions. For example, a patchy environment might give selective advantage to different alleles in different parts of the population range. **Balanced polymorphism** occurs when natural selection maintains diversity at a gene locus. Sometimes it is the heterozygote that has a selective advantage. This is called **heterozygote advantage**, and we have seen examples of it already, with the sickle-cell recessive allele conferring an advantage to the heterozygote in resisting malaria, and the cystic fibrosis recessive allele conferring an advantage to the heterozygote in surviving cholera. Another example of balanced polymorphism is **frequency-dependent selection**, where natural selection preserves a balance in the relative abundance of different alleles in a population.

Natural selection results in **adaptive evolution**. Through generations, populations become better adapted to their environments. We have spoken of an organism's **fitness**, of the relative contribution that an organism makes to the gene pool of the next generation. This is Darwinian fitness. The measure of fitness is a relative one, comparing the contributions of different genotypes to the gene pool. If one genotype leaves on average only 75% of the offspring of another, we speak of a **relative fitness** of 0.75.

Natural selection acts on phenotypes, not genotypes. Therefore, selection is indirect; it selects for favorable genotypes by acting on phenotypes. Most phenotypic characters are complex in their inheritance; they are influenced by a variety of genes as well as by the environment. The inheritance of such characters is called **multigenic** (Chapter 12). Also, a gene can influence more than one character, something we have called **pleiotropy** (Chapter 12). A mutation in a pleiotropic gene might enhance some characteristics but leave others worse. Whether such a mutant allele will be selected for depends on the balance between the positive and negative effects. Aging may be explained in part by pleiotropic genes, which may have beneficial effects early in life, but be detrimental later in life.

Natural selection can affect traits in different ways. **Stabilizing selection** selects against extremes for a trait, and will cause a population to more narrowly clump around an average value for a character or trait. That average value would be the one with the highest selective advantage for the character. **Directional selection** shifts the frequency distribution for a trait in one direction with passing generations. This might occur if organisms were reacting to a change in the environment. We saw an example of this kind of selection with pepper moths as tree bark darkened, and darker morphs had a selective advantage. **Diversifying selection** selects for variants at the extremes for a phenotypic trait, selecting against those with a middle value. A patchy environment might contribute to diversifying selection.

Sexual selection refers to mating preferences that result in some genotypes gaining in representation in the gene pool. The selection is based on secondary sexual characteristics, which enhance reproductive success. These characteristics may not be of any use other than increasing the likelihood of mating. Colorful plumage in birds is one example of such characters. Antlers in mammals are another, but these play more than one role, as they contribute to the animal's defense as well as being used for display and mate attraction. The peacock's large tail feathers, on the other hand, may actually detract from survival of the individual, but if the advantage for reproduction is high enough, they will be selected for anyway.

Natural selection cannot make perfect creatures for the following reasons:

1. Environments change.

2. Organisms are locked into historical constraints. Mutational changes will modify preexisting structures from ancestors rather than build an entirely new organism from scratch. Human aggression may be an example of such an historical constraint acting in, and limiting, society today.

3. Adaptations often are compromises. An example of such a compromise is structural rigidity versus flexibility. Human arms must be rigid enough to support weight, but

increased flexibility certainly would be beneficial when you try to retrieve an object dropped behind a heavy couch or in an engine compartment.

4. Chance plays a role in evolution. Such events as genetic drift influence what we are, and so do chance mutations. Even the most fit individual might be a meal for a predator.

5. Organisms are limited by the heritable variations that exist in the population or species. Natural selection can only operate on existing genetic variation, and genetic variation in any population is limited.

Nevertheless, natural selection has produced wonderful adaptations and exquisite control mechanisms.

Topic Test 2: Genetic Variation in Populations

True/False

1. Natural selection always chooses the best genotype, and so always reduces genetic variation.

2. If a population is polymorphic, it will have more than one allele at the corresponding gene locus.

3. Mutations in pleiotropic genes can help some characteristics of an organism while damaging others.

Multiple Choice

4. In a population of West African finches, two bill sizes are found. One is large for large seeds; the other is small for small seeds. No medium-sized bills are present in the population. Which of the following has probably occurred?
   a. Divergent selection
   b. Directional selection
   c. Stabilizing selection
   d. Sexual selection

5. An example of frequency-dependent selection is
   a. when cystic fibrosis genes are selected for in a population.
   b. whenever Hardy-Weinberg equilibrium holds in a population.
   c. the bottleneck effect.
   d. gene flow from one population to another.
   e. when one species mimics several other species, and the number of mimics of each type is a basis for selection.

Short Answer

*6. How quickly will natural selection remove a lethal mutation from a population?

Topic Test 2: Answers

1. **False.** Several instances where natural selection contributes to genetic variation are described in the text. Balanced polymorphisms and divergent selection are examples.

2. **True.** This is the definition of polymorphic.

3. **True.** It is quite possible for this trade-off to occur, as discussed in the text.

4. **a.** Divergent selection is a selection for extremes. In this case, large and small bills could have been generated by divergent selection acting on an intermediate bill size.

5. **e.** This is the only answer in which the relative abundances (frequencies) of different alleles are a basis for natural selection (frequency-dependent selection).

6. **It depends on the mutant.** A dominant mutant that kills before reproduction will disappear from a population in one generation. It will only appear again due to a new mutation. A dominant mutant that kills later in life, such as Huntington's disease, can remain in a population for many generations, depending on whether it has any effects early in life. If it is advantageous early in life, it could, on balance, even be selected for by natural selection! A lethal, recessive mutation also can remain in a population for a long period of time in heterozygous form, and can even be maintained by natural selection if it confers a great enough benefit in heterozygous form, as we have seen for cystic fibrosis and sickle-cell anemia.

APPLICATION: IMPACT OF BOTTLENECKS ON THE CHEETAH

The cheetah is a remarkable creature. In the cheetah, evolution has shaped a running machine, one that is aerodynamic and designed for high-speed chases. This sleek feline is the fastest land animal, said to be able to reach speeds of 60 to 70 mph for brief periods. It uses its speed to catch prey.

Although not without controversy, it appears that cheetah populations have relatively little in the way of genetic diversity compared to other carnivores. There is little heterozygosity in tested populations of cheetahs. Cheetah populations appear to have fixed alleles at most gene loci.

What caused this lack of genetic diversity? Cheetahs are believed to have undergone a population bottleneck during an ice age 10,000 years ago. As cheetah numbers declined, so apparently did their genetic diversity. A second bottleneck occurred more recently as a result of human hunting and destruction of habitat, probably resulting in a further reduction in genetic diversity.

A current debate concerns whether the lack of genetic diversity among present populations of cheetahs places them at an increased risk of extinction. For example, some argue that the lack of diversity makes the populations more vulnerable to infections. With more genetic diversity, a subpopulation might more likely survive a deadly virus. It is clear that the lack of genetic diversity leaves evolution little to work with. Remember that natural selection can act only when different alleles are present in a population. Certainly the lack of genetic diversity will not help cheetah populations adjust to environmental changes. While the discussions about the relationship between genetic diversity and risk of extinction continue, it is clear to all that the greatest threat to the future survival of cheetahs probably is the continued human encroachment on cheetah habitat.

Chapter Test

True/False

1. Heterozygote advantage and frequency-dependent selection are examples of balanced polymorphism.

2. Hardy-Weinberg equilibrium requires random mating and isolation from other populations.

3. Assortative mating and inbreeding are examples of nonrandom mating.

Multiple Choice

4. A population in Hardy-Weinberg equilibrium contains 9% homozygous dominant individuals. What percentage of individuals are homozygous recessive in the population?
   a. 9%
   b. 91%
   c. 82.81%
   d. 81%
   e. 49%

5. A population that is NOT in Hardy-Weinberg equilibrium contains 3% homozygous dominant individuals and 30% homozygous recessive individuals. What percentage of individuals are heterozygous?
   a. 3%
   b. 1.8%
   c. 18%
   d. 67%
   e. Cannot tell from the information given.

6. Sexual selection
   a. results in more females being born than males.
   b. results in more males being born than females.
   c. results from secondary sex characteristics influencing the likelihood of mating.
   d. is not a form of natural selection.
   e. results in either a or b.

7. Genetic drift can be caused by
   a. bottleneck.
   b. founder effect.
   c. mutations.
   d. both a and b.
   e. all of the above.

Short Answer

8. At a particular gene locus, a population in Hardy-Weinberg equilibrium has 16% homozygous recessive individuals. What fraction of the gene pool contains dominant genes at this locus?

Chapter Test: Answers

1. **T** 2. **T** 3. **T** 4. **e** 5. **d** 6. **c** 7. **d**

8. Since $q^2 = 0.16$, $q = 0.40$. Then $p = 1 - q = 0.60$, or 60%.

Check Your Performance:

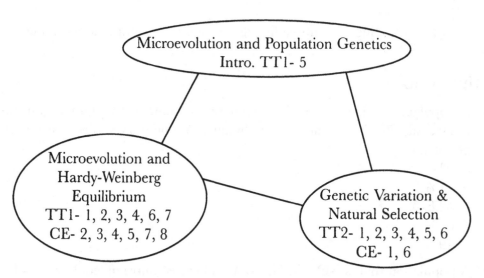

Key: TT = Topic Test; CE = Chapter Exam. Numbers indicate exam questions. Some questions are listed more than once if they refer to more than one topic.

UNIT IV:
ANIMAL PHYSIOLOGY

CHAPTER 18

General Principles of Physiology

Anatomy is the study of structure; physiology is the study of function. Structure and function are closely related to one another. The next several chapters emphasize the physiology of several different organ systems. The animal systems selected are among the more difficult ones to understand.

In earlier chapters we learned about atoms and how they make up molecules, and about molecules and how they make up cells. This chapter contains an overview of tissues, made up of cells; organs; and selected organ systems in animals. It also introduces some major principles in physiology.

ESSENTIAL BACKGROUND

- **Elementary geometry**
- **Negative feedback (Chapter 6)**

TOPIC 1: TISSUES, ORGAN SYSTEMS, AND SURFACE AREA

KEY POINTS

✓ *What are the four principal types of tissues found in animals?*

✓ *What are the principal organ systems in vertebrates?*

✓ *What important relationship between surface area and volume impacts greatly on the structure and function of living organisms?*

✓ *What special adaptations have been used to increase surface area in different cells, tissues, and organs?*

Tissues are groups of cells with a common structure and function. There are four principal types of tissues in animals. **Epithelial tissue** is found on the surface. It not only forms the skin but also forms the surface of organs and cavities within the body. A variety of different cells and shapes of cells comprise epithelial tissue.

Connective tissue is diverse. It is distributed throughout extracellular matrix, and usually generates and functions within the matrix. Bone is a form of connective tissue, and so is blood. Tendons and ligaments also are examples, and contain collagen, which is made of protein fibers consisting of a triple helix of polypeptide chains. Collagen fibers are composed of bundles of macrofibrils, which in turn consist of microfibrils, all of which are collagen protein. Collagen also is found in skin, bone, cartilage, and blood vessels. Adipose tissue, containing cells that store fat, is a form of connective tissue as well. Adipose cells pad and insulate the body and store fat as fuel.

Muscle tissue includes skeletal muscles, which underlie our voluntary movements. Smooth muscle is associated with internal organs and is not normally under voluntary control. Heart muscle forms our pump for circulating blood. Muscles are the most abundant tissue type in most animals, accounting for two-thirds of the mass of a human in good physical shape.

The fourth major type of tissue is **nervous tissue**, which contains neurons and glia. We will discuss how neurons function in a later chapter.

Tissues are grouped to form organs, and organs function in teams called organ systems. The principal **organ systems** in vertebrates are as follows:

- Circulatory or cardiovascular (transport): heart, blood vessels, blood
- Digestive (digests food): mouth, stomach, intestines, and so on
- Endocrine (hormonal coordination): pituitary, adrenal, thyroid, and so on
- Excretory (urinary; waste removal): kidney, bladder
- Immune (body defense): lymphocytes, macrophages, antibodies, and so on
- Integumentary (protects body): skin, nails, hair, and so on
- Muscular (movement): skeletal, smooth, and cardiac muscle
- Nervous (integration and control): nerves, spinal cord, brain, and so on
- Reproductive (reproduction): testes, ovaries, and so on
- Respiratory (gas exchange): lungs, trachea, air passageways
- Skeletal (support and protection): bones, cartilage, ligaments

A principal issue that relates structure and function concerns the effects of size on organisms. Why are cells the size they are? Why aren't cells larger? Humans consist of trillions of cells. Why so many? What limits the size of cells and animals?

One limiting factor is the ratio of surface area to volume. If a cell or body is roughly spherical, then as it increases in size, the volume will increase as the cube of the radius, while the surface area will only increase as the square of the radius. Thus, **surface area/volume $\cong$ 1/radius**. The ratio decreases as the volume of a cell increases. The amount of surface area per volume of a cell is important because cells must use their surface for exchange of gases, nutrients, and waste products. Getting oxygen to the middle of a cell requires that the oxygen penetrate the surface and then diffuse to the middle. Diffusion takes time. Hence, there is a limit to how large a spherical cell can be and still maintain an adequate supply of oxygen. The amount of surface area supplying a given volume becomes a limiting factor as cells get larger. Cells divide or enhance their surface areas by taking on shapes other than spherical. Some cells are flat, enhancing surface area and decreasing distance to the middle of the cell; others have folds and cavities that increase surface area for a given volume.

The same issue of surface area and volume repeats itself at the level of tissues and organs. The need for a large surface area contributes to the structures of certain organs. Most larger animals have extensive folds in some of their organs that provide increased surface areas for the interchange of materials. Human lungs have internal membranes with millions of microscopic air chambers, giving humans a total lung surface area of about 100 square meters, the size of a tennis court! Our kidneys contain about 8 million microscopic tubules, specialized to filter waste products from blood. Our intestines are 6 meters long, with a surface area, due to indentations, equal to that of a baseball diamond.

Topic Test 1: Tissues, Organ Systems, and Surface Area

True/False

1. Bone is a form of connective tissue.

2. Adipose tissue is a form of epithelial tissue.

3. The immune system includes skin, hair, and nails.

Multiple Choice

4. What happens as the volume of a sphere increases?
   a. The surface area decreases.
   b. The ratio of radius to diameter changes.
   c. The ratio of surface area to volume increases.
   d. The ratio of surface area to volume decreases.
   e. The volume increases as the square of the radius increases.

Short Answer

5. Consider two spherical cells. One has a radius of 1 micron; the other has a radius of 2 microns. Compare the ratio of surface area to volume of the large cell to that of the smaller cell.

Topic Test 1: Answers

1. **True.** Bone is one of many different kinds of connective tissue.

2. **False.** Adipose tissue is another form of connective tissue.

3. **False.** The integumentary system includes what is listed. The immune system includes lymphocytes, macrophages, antibodies, and so on.

4. **d.** As the radius increases, the surface area increases, but not as much as the volume does.

5. Using the formula given, that surface area and volume are proportional to 1/radius, the larger cell has a surface area–volume ratio that is one-half that of the smaller cell.

TOPIC 2: HOMEOSTASIS AND TEMPERATURE REGULATION

KEY POINTS

✓ *What is homeostasis?*

✓ *How is body temperature regulated in humans?*

We have been programmed through evolution by natural selection to maintain a rather constant internal environment. We have physiological control systems that regulate body temperature, blood pH levels, blood Ca^{2+} levels, blood glucose levels, body weight, and water balance, among others. Maintaining this steady state internally is called **homeostasis**. Our physiological systems maintain so-called **set points** for variables, such as body temperature, that ensure smaller internal fluctuations in response to large external ones. Such controls can be very important for survival. For instance, the pH of blood does not usually vary more than a tenth of a pH unit from 7.4. If it is a few tenths off, the individual is very sick.

We use **feedback** systems to maintain these set points. The form of feedback is **negative feedback** because a change in the value of one of the variables causes a reaction that reverses the change back toward the set point. We first examined negative feedback in allosteric enzymes in biochemical pathways, where it was used to maintain certain concentrations of products in cells (Chapter 6). Here we use the control of body temperature as an example of a homeostatic mechanism using negative feedback.

Human body temperature normally is maintained at a set point of just under 37°C. This is called **thermoregulation**. We have a control system that functions rather like the thermostat that maintains temperature in a room or house. Our body's control system requires a **receptor**, which measures the variable; a **control center**, which activates the negative feedback changes; and **effectors**, which bring about the changes. There is a temperature receptor in the hypothalamus of the brain, and it is linked to a hypothalamic thermostat (control center) which controls body responses (effectors) ranging from sweating to shivering.

If the temperature of the hypothalamus increases above its set point, we begin to sweat, blood will move toward the body surface, and we usually move out of the sun or reduce physical activity levels. The feedback of a change in temperature at the hypothalamus induces changes in the body and in behavior that tend to reverse the temperature change back toward the set point.

If our body temperature decreases below the set point, skin blood vessels constrict, reducing heat loss, and skeletal muscles are activated, producing shivering, which generates heat. These effects tend to bring body temperature back toward the set point.

The set points can change as well. Infections, such as a flu, can cause the set point for body temperature to increase. If the increase is rapid, we feel cold and shiver, as our body temperature increases by thermoregulation to match the new set point. When our fever "breaks," the set point has dropped, and our body temperature is temporarily higher than the new, lower set point, resulting in sweating.

Much of physiology is the study of homeostasis and the feedback systems that have evolved to maintain set points. Understanding the mechanisms of homeostasis also becomes the foundation for understanding what goes wrong during illness and disease.

Topic Test 2: Homeostasis and Temperature Regulation

True/False

1. Set points lock the value of a variable so that it never deviates from the set point.

2. Set points themselves are fixed and never change.

3. During a fever, the body's temperature rises above the set point.

Multiple Choice

4. The set point for body temperature increases. Because of thermoregulation, one expects
   a. the individual will begin to sweat, and blood vessels in the skin will dilate (open).
   b. the individual will shiver and blood vessels in the skin will contract.
   c. the hypothalamus will rapidly return the set point to normal.
   d. the set point will have little effect on body temperature.

Short Answer

5. Briefly define homeostasis and give an example.

Topic Test 2: Answers

1. **False.** Values can deviate from a set point. Homeostatic systems return the values back toward the set point.

2. **False.** Consider the change in set point that occurs with a fever.

3. **False.** The set point rises, and then the body's temperature rises to match the new set point.

4. **b.** The body's temperature is below the set point, and so the individual responds by raising the body's temperature.

5. Homeostasis refers to a living organism's maintaining a relatively constant internal environment. Variables are kept within a narrow range by control systems that use negative feedback to regulate the variables, even in the presence of significant environmental variation. Examples in humans include body temperature, blood glucose (sugar) levels, and blood calcium levels.

APPLICATION: HEAT STROKE

Under extreme conditions, the attempt of the body to maintain a body temperature near set point can lead to other problems. Consider an individual who is exercising on a hot day. Exercise requires the production of ATP by mitochondria, and the ATP is hydrolyzed during muscle contraction. Both production and hydrolysis of ATP generate considerable heat, which can increase body temperature above the set point by several degrees.

The increase in body temperature during exercise triggers hypothalamic temperature regulation, causing sweating, which helps to control the rise in body temperature through evaporative cooling, but such sweating also causes a loss of body fluids. Sweat is primarily water, with some salt and organic compounds. Without proper fluid replacement, prolonged exercise can lead to dehydration and this can produce overheating. In high humidity, the problem can be exacerbated because sweating is less effective—dripping sweat does not contribute to evaporative cooling. While sweating can be more effective in dry conditions, the loss of body fluid may not be as noticeable. In both cases adequate fluid replacement is very important. During exercise, the body can lose water at the rate of 1 to 2 pounds per hour or faster. Each pound is about a pint of fluid. Without fluid replacement, water begins to be removed from blood plasma, further reducing the ability of the body to control temperature. At extreme body temperatures, the ability of the hypothalamus to control temperature also is reduced as the hypothalamus itself heats up.

The body does acclimate, over a period of 1 to 2 weeks, to high heat and humidity. But even when acclimated, fluid consumption is very necessary to offset losses. **Heat stroke** is an extreme state that can result from insufficient fluid replacement, or from exposure to extreme environmental temperatures for prolonged periods. The body actually ceases, or nearly ceases, sweating as body fluids decline in volume. The skin appears dry to the touch. The temperature of the individual rises to dangerous levels, and coma, brain damage, and death can result. Individuals usually are treated by sponging their body with alcohol or bathing in ice water, in an attempt to bring body temperature back under control.

Chapter Test

True/False

1. The excretory system includes the kidney and bladder.

2. Collagen fibers are abundant in certain forms of connective tissue.

3. As a balloon expands, its surface area increases faster than its volume.

Multiple Choice

4. Concerning body temperature control in humans,
   a. the control system for body temperature is located in the hypothalamus of the brain.
   b. the set point for body temperature can change, as when one gets a fever.
   c. the set point for body temperature is constant, and does not change for an individual.
   d. both a and b are correct.
   e. both a and c are correct.

5. As the volume of a sphere grows, the surface area–volume ratio change is proportional to
   a. the radius.
   b. the radius squared.
   c. 1/radius.
   d. 1/(radius squared).
   e. none of the above.

Short Answer

6. What are the three different kinds of muscles found in humans?

Chapter Test: Answers

1. **T** 2. **T** 3. **F** 4. **d** 5. **c**

6. The three kinds of muscle are skeletal, smooth, and heart muscle.

Check Your Performance:

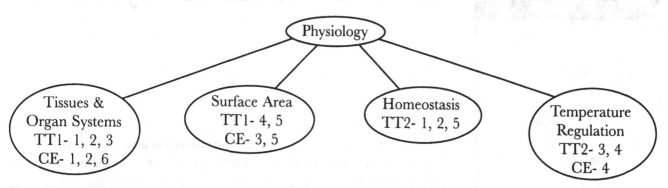

Key: TT = Topic Test; CE = Chapter Exam. Numbers indicate exam questions. Some questions are listed more than once if they refer to more than one topic.

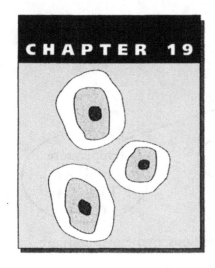

CHAPTER 19

Cardiovascular System

The circulatory system is called the **cardiovascular system** in vertebrates and is responsible for delivering nutrients and oxygen to interstitial fluid while removing waste materials and carbon dioxide. **Interstitial fluid** directly bathes our cells, which then gain nutrients and oxygen from the interstitial fluid. In this chapter we examine the three components of the cardiovascular system—heart, blood vessels, and blood.

ESSENTIAL BACKGROUND

- Gap junctions (Chapter 5)
- Endocytosis and exocytosis (Chapter 4)

TOPIC 1: HEART

KEY POINTS

✓ *Why is the heart considered to be two pumps in one?*

✓ *What is the sequence of flow of blood through the heart?*

✓ *What pathway does blood take as it flows from the heart to and from the lungs?*

The human heart must pump blood continuously throughout our lives. There is no "time-out" for repairs, which must be done while the heart is pumping. Even at 60 beats per minute (bpm), which is a resting heart rate for many humans, the heart will contract over 2 billion times in a lifetime.

Vertebrates have closed circulatory systems—the blood circulates within a closed system of heart and blood vessels. Some invertebrates have open circulatory systems. Not all vertebrates have the same number of heart chambers. Some, such as fish, have two heart chambers; amphibians have three chambers; and birds and mammals have four chambers, with two **atria**, each leading into a separate ventricle. The two **ventricles** are the larger chambers that pump blood from the heart.

Deoxygenated blood from the body returns in veins to the right atrium, which contracts to help fill the right ventricle. The **right ventricle** pumps deoxygenated blood from the heart to the lungs by way of the **pulmonary artery**. Oxygenated blood returns from the lungs to the left atrium by way of the **pulmonary vein**. Then the left atrium contracts to fill the **left ventricle**, which pumps the blood from the heart to the rest of the body.

Your heart is only about the size of your fist. The wall of the heart is mostly cardiac muscle tissue. The ventricles have thick walls and are more powerful than the atria.

The **heart cycle** is the sequence of events that occurs during a heart beat. There are two phases: **systole**, when the muscle contracts and blood is pumped, and **diastole**, when the ventricles are filling. The atria contract just before the ventricles, and complete the final 30% of filling of the ventricles.

Atrioventricular **heart valves** close to prevent backflow of blood from the ventricles into the atria when the ventricles contract. Semilunar heart valves prevent blood from flowing back from the arteries to the heart as the ventricles are filling. The closing of these valves causes most of what one hears when listening to a heart beat with a stethoscope.

In a typical adult, the resting heart rate is about 60 to 70 bpm. In an athlete this rate can be as low as 30 bpm! **Cardiac output** is the heart rate (bpm) times the stroke volume (the volume of blood that exits the left ventricle with each stroke). Thus, cardiac output is the amount of blood pumped by the heart per minute (units = volume/minute). A typical stroke volume is 75 milliliters. From this, you can calculate that cardiac output is about equal to the total volume of blood in the entire body (about 5 liters) pumped per minute. Cardiac output can increase fivefold during exercise.

Heart muscle cells have the special property of contracting spontaneously. The heart is designed to coordinate this spontaneous contraction of each muscle cell. The rate of contraction is set by a specialized **pacemaker** region called the **sinoatrial (SA) node**. It is located in the wall of the right atrium, and self-generates electrical potentials that cause contraction. The muscle cells of the heart are connected by gap junctions (Chapter 5), which allow the spread of excitation and contraction. Another patch of nodal tissue, the **atrioventricular (AV) node**, is at the bottom center of the atria, and is responsible for spreading the wave of excitation through the ventricles, causing them to contract.

While the contraction rhythm is set by the SA node, and the SA node can "fire" automatically, it also is influenced directly by two sets of nerves. One set speeds up and the other slows down the heart rate. Hormones such as adrenaline (epinephrine) also can increase the heart rate, as will be discussed in the next chapter. An increased load of returning venous blood also increases the heart rate, which is one way exercise triggers an increase in cardiac output.

The wave of electrical excitation that spreads across the heart and causes muscle contraction can be measured by electrodes placed on the chest, and this measurement is called the **electrocardiogram (EKG** or **ECG)**.

Topic Test 1: Heart

True/False

1. If cardiac output doubles with no change in stroke volume, then resistance in arteries must have decreased.

2. The pulmonary vein contains deoxygenated blood.

3. Heart valves prevent blood backflow and cause heart sounds.

Multiple Choice

4. A heart is beating twice each second. Its stroke volume is 100 milliliters. The cardiac output is (note: 1,000 milliliters = 1 liter)
   a. 200 milliliters.
   b. 2 liters/minute.
   c. 12 liters/minute.
   d. 6 liters/minute.
   e. none of the above.

5. The pulmonary artery in mammals carries blood
   a. from the left ventricle of the heart to the lungs.
   b. from the left ventricle to the left lobe of the lungs, and from the right ventricle to the right lobe of the lungs.
   c. from the organ systems of the body to the right atrium.
   d. from the lungs to the right ventricle.
   e. from the right ventricle to the lungs.

Short Answer

6. Briefly describe the flow of blood from the right atrium to the left ventricle in humans. What structures does the blood pass through?

Topic Test 1: Answers

1. **False.** The heart rate must have doubled.

2. **False.** The pulmonary vein carries oxygenated blood from the lungs to the left atrium of the heart.

3. **True.** Heart valves shut automatically by pressure on one side of them, but not the other, and thereby prevent backflow.

4. **c.** 2 beats/second = 120 bpm. Cardiac output = 120 bpm × 100 milliliters = 12,000 milliliters/minute = 12 liters/minute.

5. **e.** The pulmonary artery carries blood away from the heart.

6. Right atrium → right ventricle → pulmonary artery → lungs → pulmonary vein → left atrium → left ventricle.

TOPIC 2: ARTERIES, CAPILLARIES, VEINS, AND BLOOD

KEY POINTS

✓ *What is the pathway for blood flow from the heart through the body?*

✓ *Compare blood pressure and velocity in the veins, arteries, and capillaries.*

✓ *How does exchange of material occur across capillary walls?*

By definition, **arteries** carry blood away from the heart, while **veins** carry blood toward the heart. Notice that not all arteries carry oxygenated blood—consider the pulmonary artery.

Blood from the left ventricle is pumped into the **aorta**, the largest artery (2.5 cm diameter), and on to the rest of the body. The first branches off of the aorta are the **coronary arteries**, which carry blood to nourish the heart muscle. Other arteries branch from the aorta to carry blood to muscles and organs throughout the body. The arteries branch into finer **arterioles**, tiny vessels that further branch into microscopic **capillaries**, which are the numerous exchange sites between the blood and the interstitial fluid that bathes cells.

Capillaries merge into larger, but still tiny, **venules**, and these converge into **veins**, which return blood to the heart. The circuit actually is not complete until the right side of the heart pumps the blood to the pulmonary artery and on to the lungs, where another set of capillaries allows for exchange of carbon dioxide and oxygen, and from the lungs the blood returns, via the pulmonary vein, to the left side of the heart. The arteries, veins, and capillaries of a human cover a distance of 100,000 kilometers, or 62,000 miles! Most of this is in the capillaries.

The velocity of blood flow is highest in the arteries, slows considerably in capillaries, and speeds up again in veins. While the diameter of each capillary is small, because there are so many of them, the total cross-sectional area and volume in the capillaries are much greater than the area and volume in either the arteries or the veins. The greater total area is the cause of the slowing of blood velocity in the capillaries.

The walls of capillaries are thin, consisting of a single layer of cells, which can be somewhat permeable, enhancing the exchange of materials between blood and interstitial fluid. Some of the exchange occurs by endocytosis on one side of the capillary wall and exocytosis on the other. Other exchange occurs by pressure, forcing materials through the clefts between cells. Finally, some material simply diffuses across the wall.

Blood pressure is much higher in arteries, and is greatest during systole, when the heart contracts. Blood pressure is determined by a combination of cardiac output and peripheral resistance to flow. The arteries are elastic, and they expand somewhat during systole. Arteries, arterioles, venules, and veins all contain smooth muscle, whose state of contraction influences blood pressure by changing the diameter of the vessels. Stress, physical or emotional, can influence the state of contraction of the blood vessels by influencing smooth muscle contraction. The smooth muscle is found in the middle of arterial and venous walls, with an endothelial layer on the inside and connective tissue on the outside. Blood pressure is lower in the capillaries, and lower still in the veins. Veins contain one-way valves to aid in the flow of blood back toward the heart.

Sets of capillaries are not always open to the flow of blood. Some main channels between arterioles and venules called **thoroughfare channels** always are open to flow. True capillaries branch off from these thoroughfare channels. Smooth muscle sphincters at the branch points can be open or closed to regulate blood flow. This allows blood to be directed to where it is needed most at any particular time. One major effect of exercise and physical fitness is thought to be an increase in the capillary beds in the exercised muscle tissue.

The volume of fluid and blood components that exit capillaries for interstitial fluid is offset by a nearly equal volume returning. There is a net loss of fluid from capillaries nearer to the arteriole side, and a net gain toward the venule side, due to pressure differences along the capillary walls. A small fraction of the fluid that flows across capillaries to interstitial fluid returns to veins via the **lymphatic system**, which is part of the immune system. This flow helps to clear the fluid of bacteria and viruses, especially as it passes through **lymph nodes**, which are filled with white blood cells, or leukocytes (see Chapter 22).

Blood contains plasma, which includes water, ions, proteins, and nutrients. Blood also contains cells, which include erythrocytes (red blood cells, for transporting oxygen), leukocytes (white blood cells, part of the immune system), and platelets (for blood clotting). For more detail on blood, see the web page.

Topic Test 2: Arteries, Capillaries, and Veins

True/False

1. Coronary arteries are the largest arteries.

2. Arterioles pass blood directly to venules.

3. Blood plasma contains proteins.

Multiple Choice

4. The lymphatic system
    a. has blood flowing directly through it.
    b. takes in interstitial fluid, filters out bacteria and viruses, and returns fluid to the veins.
    c. is situated inside of arteries.
    d. contains lymph nodes containing white blood cells.
    e. includes both b and d.

5. A red blood cell is in an artery in the left arm. How many capillaries must this cell pass through before it is returned to the left atrium of the heart?
    a. One
    b. Two
    c. Three
    d. Four
    e. None

6. Concerning veins,
    a. blood pressure is higher in veins than in capillaries.
    b. all veins carry blood that is low in oxygen.
    c. veins deliver blood to capillaries.
    d. blood flow in veins is away from the heart.
    e. veins contain valves that help to prevent backflow of blood.

Short Answer

7. In the mammalian cardiovascular system, blood flows from the left atrium to the _____ and then to the aorta. It then passes through _____ (how many?) capillary bed(s) before again returning to the left atrium. In capillary beds, the speed of blood flow is _____ (greater than, less than, the same as) the speed of blood flow in the aorta. In veins, the speed of blood flow is _____ (greater than, less than, the same as) the speed in capillaries.

Topic Test 2: Answers

1. **False.** The aorta is the largest artery. Coronary arteries branch off of the aorta.

2. **False.** Arterioles pass blood to capillaries.

3. **True.** Those proteins include pH buffers, blood-clotting fibrinogen, and antibodies.

4. **e.** The lymph system takes in interstitial fluid and contains lymph nodes with white blood cells.

5. **b.** It will pass through one capillary in the arm, in order to get to a vein in the arm, but then must pass through a second capillary in the lungs before getting to the left side of the heart.

6. **e.** Such valves are found in veins, but not in arteries.

7. Left ventricle; two (one in organs or muscles, one in the lung); less than; greater than

APPLICATION: CARDIOVASCULAR DISEASE

Cardiovascular disease is the number one killer in the United States today and can cause heart attacks and strokes. Many of these are caused by blocking of the arteries. Especially important are the coronary arteries, which provide nourishment and oxygen to the heart. A blood clot can obstruct the flow, and is more likely to occur in an artery that has been narrowed by **atherosclerosis**, which consists of growths or **plaques** that develop on the inner walls of arteries and narrow the bore of the vessels. One form of atherosclerosis is **arteriosclerosis**, or hardening of the arteries, which occurs when calcium deposits harden the plaques. One sign of artery narrowing is **hypertension**, or high blood pressure, caused by the increased resistance of blood flow in clogged arteries.

The risk factors for circulatory disorders such as atherosclerosis and hypertension include heredity, smoking, lack of exercise, stress, a diet rich in fats, and high cholesterol levels in the blood, especially if such cholesterol is associated with **low-density lipoproteins (LDLs)** in the blood. Such LDLs consist of many molecules of cholesterol and other lipids bound to protein. High levels of LDLs are associated with higher degrees of plaque formation in arteries. In contrast, **high-density lipoproteins (HDLs)** actually appear to be good, and exercise increases HDL levels, while smoking reduces it. Epidemiological studies indicate that smoking causes more deaths from cardiovascular disease than from lung cancer.

The number of deaths from cardiovascular diseases actually has been declining, and is down 25% from the peak in the 1970s. Lifestyle is very important in reducing the risk of premature death and disability from heart attacks and strokes. Exercise and diet are very important, and not smoking appears to be even more important. There is little one can do, short of stepping in front of moving trains and the like, to increase the risk of premature death and disability more than by smoking.

Chapter Test

True/False

1. A high concentration of HDL can be good for you.

2. Stress can influence the state of contraction of blood vessels.

3. Blood contains both plasma and cells.

Multiple Choice

4. In capillaries
   a. blood pressure is greater than elsewhere in the circulatory system because each capillary has such a small diameter.
   b. velocity of blood flow is greater than elsewhere in the circulatory system.
   c. the total cross-sectional area for flow of blood is greater than elsewhere in the circulatory system.
   d. None of the above is correct.
   e. All of the above are correct.

5. In vertebrates, where is blood pressure the greatest?
   a. In veins
   b. In arteries
   c. In capillaries
   d. It is equal everywhere in the circulatory system.

6. In the human heart, which chamber is responsible for pumping blood to the lungs?
   a. The left ventricle
   b. The right ventricle
   c. The left atrium
   d. The right atrium
   e. None of the above.

Chapter Test: Answers

1. **T**   2. **T**   3. **T**   4. **c**   5. **b**   6. **b**

Check Your Performance:

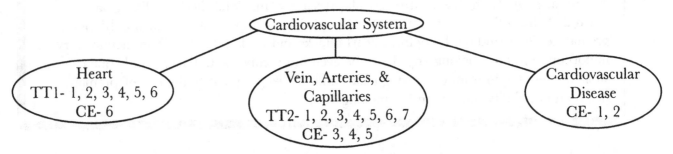

Key: TT = Topic Test; CE = Chapter Exam. Numbers indicate exam questions. Some questions are listed more than once if they refer to more than one topic.

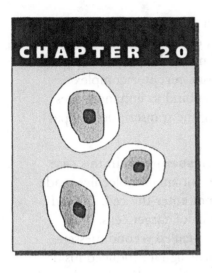

CHAPTER 20

Endocrine System and Hormones

Two major, interrelated communication systems have evolved in animals. One is the nervous system, which we will consider in the next chapter. The other is the endocrine system. The endocrine system conveys signals between cells in the body. **Hormones** are the chemical signals that convey messages via the circulatory system. The hormones are made by **endocrine cells**, which usually are assembled into **endocrine glands**.

We identify the major endocrine glands and explore how hormones are used in the regulation and control of homeostasis. Understanding the endocrine system also serves as a foundation for understanding many medical disorders, and the endocrine system offers numerous target sites for medications.

Other processes make use of chemical signals other than hormones.

ESSENTIAL BACKGROUND

- **Membranes (Chapter 4)**
- **Transcription and enhancers (Chapters 10 and 15)**

TOPIC 1: MECHANISMS OF HORMONE ACTION

KEY POINTS

✓ *How do hormones influence target cells?*

✓ *What are the two general mechanisms of hormone action?*

✓ *What are second messengers and G proteins?*

Hormones are secreted into the blood and are carried throughout the body. Only certain types of cells respond to a given hormone. These are called the **target cells** for that hormone, and they contain hormone **receptors**, which bind to the hormone and trigger the cell's response. The binding of a hormone to its receptor is another example of the key-and-lock fit that we discussed with substrates and enzymes. The fit of the hormone "key" to the receptor protein "lock" can trigger a cascade of events in the cell. The end result in the target cell depends on the particular kind of receptor and the machinery in the cell that is present for activation by the receptor-hormone complex.

There are two general mechanisms of hormone action, depending on the kind of hormone. One mechanism involves **steroid hormones**, which are made from cholesterol. Because they are hydrophobic, steroid hormones are able to penetrate the membrane of the target cell. Inside the cell the steroid hormones bind to receptors, and the hormone-receptor complex becomes a transcription factor, or activates transcription factors. These bind to enhancer sites (Chapter 15), or other regulatory sites on the DNA, thereby regulating the transcription of particular genes.

The other general mechanism of hormone action involves **second messengers**. In this case the hormones are either peptide hormones, consisting of polypeptides, or small molecules made by chemically modifying common amino acids. These hormones do not enter the cell. Instead they bind to membrane-bound receptors that are present on the surface of target cells. The receptor-hormone complex then triggers changes in the cell through so-called second messengers. The hormones can be considered the first messenger; the chemical signal produced inside the cell by the presence of a hormone on the receptor is the second messenger. For these second-messenger systems, there are **signal transduction pathways** that bring about changes in the cell in response to the hormone signal. Here we consider two known second messengers: **cyclic AMP** and **inositol triphosphate (IP$_3$)**.

Cyclic AMP is formed from ATP by the enzyme **adenylyl cyclase**. It is broken down by the enzyme **phosphodiesterase**. The receptor-hormone complex stimulates adenylyl cyclase indirectly. A class of membrane proteins called **G proteins** are stimulated to hydrolyze a GTP molecule when they contact the hormone-receptor complex. There are different kinds of G proteins. Some stimulate adenylyl cyclase, while others inhibit it. When adenylyl cyclase is stimulated, the cyclic AMP that is produced can trigger an amplifying cascade of reactions, which produce a variety of responses (enzyme activations and inhibitions, membrane permeability changes, secretions, transcription, etc.). The cascades usually involve **protein kinases**, which act by adding phosphate groups to other proteins. Those proteins are activated or inhibited by the phosphorylation, and as enzymes, can multiply the effect of a few hormone molecules. For example, in a liver cell, a few molecules of the hormone norepinephrine can cause the release of millions of glucose molecules.

When **IP$_3$** is the second messenger, the receptor-hormone complex in the membrane also activates G proteins. The G proteins, in turn, activate **phospholipase C**, the enzyme that generates IP$_3$ and diacylglycerol from a membrane phospholipid. The IP$_3$ triggers the release of calcium ions from the endoplasmic reticulum, and the calcium ions can directly trigger responses or can bind to a protein, **calmodulin**. The calcium-calmodulin complex can bind to, and activate, enzymes. The diacylglycerol that also was produced by the phospholipase C activates a membrane-bound **protein kinase C**, which acts by phosphorylating specific proteins. As with cyclic AMP, a whole host of proteins can have their activities changed by the hormone.

Topic Test 1: Mechanisms of Hormone Action

True/False

1. Endocrine cells produce hormones.

2. Protein kinase C produces IP$_3$.

3. Steroid hormones act by modifying transcription.

Multiple Choice

4. Concerning the mechanisms of hormone action, G proteins
   a. are what steroid hormones bind to.
   b. are found in the membranes of target cells that use second messengers.
   c. are what peptide hormones bind to.
   d. are what small hormones, such as norepinephrine, bind to.
   e. are all of the above.

5. Which of the following is NOT involved in the cyclic AMP second-messenger system?
   a. Adenylyl cyclase
   b. Steroid hormones
   c. G proteins
   d. Receptor proteins
   e. Protein kinases

6. Calcium release is a standard part of the response system for which of the following second messengers?
   a. A hormone
   b. Cyclic AMP
   c. IP_3
   d. Adenylyl cyclase

Short Answer

7. What three different effectors bring about responses in target cells when IP_3 is a second messenger?

8. How do steroid hormones act?

Topic Test 1: Answers

1. **True.** Endocrine cells produce and secrete hormones. They usually are found in endocrine glands.

2. **False.** Phospholipase C produces the second messenger IP_3.

3. **True.** They can pass directly into cells and bind to internal receptors that then become transcription factors, regulating gene expression.

4. **b.** The G proteins are activated by linking to receptor-hormone complexes.

5. **b.** Steroid hormones do not use second messengers.

6. **c.** The calcium then acts directly, or binds to calmodulin.

7. Calcium, calcium-calmodulin complex, and protein kinase C are the three different effectors that bring about responses in target cells.

8. Steroid hormones act by binding to receptors in target cells, thereby producing, or stimulating, transcription factors.

TOPIC 2: ENDOCRINE GLANDS AND THEIR CONTROL IN VERTEBRATES

KEY POINTS

✓ *What is the role of the hypothalamus and pituitary in the endocrine system?*

✓ *What are tropic hormones?*

✓ *What are some of the chief endocrine glands, and what do they make?*

There are more than 50 identified hormones in humans. Hormones are often involved in feedback inhibition processes that contribute to homeostasis. As such, there can be pairs of **antagonistic** hormones that act to bring about opposite changes. We examine several such pairs in the next section. Some hormones have tissues that secrete other hormones as their targets. These are called **tropic hormones**.

The **hypothalamus** is a key brain center coordinating and regulating many different hormones and endocrine glands. It uses tropic hormones to regulate the release of hormones from the **pituitary gland**, which sits just beneath it in the brain. The hormones secreted from the pituitary gland influence many other endocrine glands.

We learned in Chapter 18 that the hypothalamus is the site of temperature regulation. With its surrounding limbic system of nervous tissue, the hypothalamus also influences hunger, thirst, and other basic body functions. The hypothalamus contains **neurosecretory cells**, which combine the electrical excitability of nerve cells with the secretory function of endocrine cells—action potentials (discussed in the next chapter) in the neurosecretory cells trigger the release of hormone from the cells.

The hypothalamus makes two hormones that are secreted into the general circulation. The neurosecretory cells that make the two hormones actually extend down into the posterior (rear) portion of the pituitary gland, and it is there that the two hormones are released. One of these hormones, **oxytocin**, is involved during reproduction, and the other, **antidiuretic hormone (ADH, vasopressin)**, acts on the kidneys to retain water in the body. ADH is part of a negative feedback loop preserving homeostasis. It is released when osmolarity sensors in the hypothalamus signal that the blood is too salty (high osmolarity). ADH acts to retain body water, and the hypothalamus also signals thirst, inducing drinking. Together these actions increase water content and reduce osmolarity back to normal.

The hypothalamus also secretes a number of tropic hormones that flow a very short distance through blood vessels to the anterior region of the pituitary gland. The pituitary gland, in response to these hormones, secretes a variety of its own hormones. Among the hormones released from the anterior pituitary are **growth hormone, follicle-stimulating hormone (FSH), luteinizing hormone (LH), thyroid-stimulating hormone (TSH), adrenocorticotropic hormone (ACTH), prolactin**, and **endorphins**.

Growth hormone is involved in governing our growth during development. Too much makes giants; too little, dwarfs. FSH and LH act on the testes and ovaries, stimulating gamete production and influencing release of sex hormones. ACTH is a tropic hormone that stimulates release of steroid hormones from the adrenal cortex. TSH acts on the **thyroid gland**, which is involved in vertebrate development and maturation as well as metabolism. We examine these endocrine glands in more detail in the next section. The hypothalamus controls the release of these pituitary hormones with one or sometimes two tropic hormones, one stimulating release and one suppressing it. Thus, the hypothalamus exerts considerable control over the rate of hormone release from the pituitary, and the pituitary hormones govern a number of endocrine glands throughout the body.

There are other major endocrine glands in the body. Among them, the **pancreas** secretes insulin and glucagon, as discussed in the next section. The **pineal gland** secretes **melatonin**, which is involved in the control of biological rhythms.

Topic Test 2: Endocrine Glands and Their Control in Vertebrates

True/False

1. A neurosecretory cell is like a nerve cell that also is a secretory cell.

2. The thyroid gland is involved in insulin secretion.

3. Too little growth hormone during development can produce dwarfism.

Multiple Choice

4. ADH (antidiuretic hormone)
   a. is synthesized in the hypothalamus.
   b. is released from the pituitary gland.
   c. acts on the kidney.
   d. causes retention of water.
   e. does all of the above.

5. The hormones released from the anterior pituitary
   a. have their release controlled by tropic hormones from the hypothalamus.
   b. govern release of hormones from several different endocrine glands.
   c. include ADH.
   d. include both a and b.
   e. include all of the above.

Short Answer

6. Why would antagonistic pairs of hormones be useful in homeostasis?

7. Describe the mechanism of action of sex hormones.

Topic Test 2: Answers

1. **True.** It combines properties of both cell types.

2. **False.** The pancreas secretes insulin.

3. **True.** We now can make growth hormone through genetic engineering and use it to treat those who produce too little of their own.

4. **e.** ADH is made in the hypothalamus and released from the posterior pituitary gland.

5. **d.** The anterior pituitary is a major site of synthesis of hormones controlling the endocrine system.

6. Maintaining a set point involves being able to respond to changes in two different directions, to values above or below the set point. One of the antagonistic pair of hormones can generate responses that bring the level of the variable back to set point when it is above the set point, and the other can generate responses that bring the level back when the value is below the set point.

7. Since sex hormones are steroid hormones, they act by regulating transcription. They enter target cells and bind to receptors, and the receptor-hormone complex regulates mRNA production for certain genes in the target cells. Notice how the mechanisms described in the first section can be applied to answer a variety of questions once one knows the kind of hormone.

TOPIC 3: EXAMPLES OF HORMONAL REGULATION

KEY POINTS

✓ *How is thyroid hormone secretion regulated?*

✓ *How do calcitonin and PTH act as antagonistic hormones?*

✓ *How do insulin and glucagon act as antagonistic hormones?*

✓ *What do the adrenal glands do?*

One endocrine gland that is controlled by the hypothalamus/pituitary is the thyroid gland, which controls metabolism in mammals. The release of two iodine-containing hormones, thyroxine (T_4) and triiodothyronine (T_3), is controlled by thyroid-stimulating hormone (TSH) from the anterior pituitary. The release of TSH from the pituitary is controlled, in turn, by thyroid-stimulating hormone–releasing hormone (TRH), from the hypothalamus.

T_4 and T_3 control metabolism in many body cells. By feedback, the thyroid hormones inhibit release of TRH from the hypothalamus, as does TSH, thus self-regulating under normal conditions. When regulation is disrupted, disorders result. Excess T_4 and T_3 levels cause **hyperthyroidism**, with symptoms including high body temperature, profuse sweating, weight loss, irritability, and high blood pressure. Too little of these hormones causes **hypothyroidism**, whose symptoms include weight gain, lethargy, and intolerance to cold.

The thyroid hormones contain iodine, and a low level of iodine in the diet can lead to enlarged necks from enlarged thyroid glands, a condition known as **goiter**, caused as the gland grows in response to the need for more of the thyroid hormones. Goiter is largely prevented today by simply adding a bit of iodine to table salt, thereby ensuring enough iodine in our diets.

The thyroid gland also contains endocrine cells that secrete **calcitonin**, which is involved in the regulation of calcium levels in the blood. If calcium concentration in the blood is above the set point, calcitonin is released by the thyroid and stimulates bone formation, which reduces calcium levels in the blood. Calcitonin is part of a homeostatic regulatory mechanism that includes **parathyroid hormone (PTH)**. PTH is made by parathyroid glands, which are located at the thyroid gland. PTH acts in an opposite direction, stimulating degradation of bone and thereby causing a release of calcium ions into blood. Here we have a good example of antagonistic hormones, each regulating calcium levels by negative feedback. Together they are able to respond to both increases and decreases in blood calcium levels. It is rather like a thermostat linked to both heater and air conditioner, and thus able to regulate in both directions.

The **pancreas** produces an antagonistic pair of hormones, **insulin** and **glucagon**, which are peptide hormones that influence blood glucose levels. Insulin acts to decrease, and glucagon increase, glucose concentration in the blood. The set point for glucose is about 90 milligrams per 100 milliliters. When blood glucose levels rise, as occurs after a meal, insulin is secreted. Most cells of the body have insulin receptors, which, when insulin binds, causes the cell to take up glucose, reducing blood glucose concentration. Insulin also acts on the liver to reduce the breakdown of glycogen and release of glucose.

If blood glucose levels fall below the set point, glucagon is released from the pancreas. The glucagon stimulates glycogen hydrolysis in the liver, which results in glucose release into the blood, returning blood glucose levels to normal. Here, again, we have antagonistic hormones, each helping to maintain normal blood glucose concentration by negative feedback.

The **adrenal glands** sit on top of the kidneys. They are like two glands in one, and both respond to stress. The **adrenal medulla**, on the inside of the gland, secretes **epinephrine** and **norepinephrine**, both made from the amino acid tyrosine, and both are involved in short-term stress. Target cells for epinephrine and norepinephrine use cyclic AMP as a second messenger. They include the heart, which increases its rate and stroke volume in response to the released hormones, and smooth muscle in blood vessels, which enlarge to increase blood flow to skeletal muscles and contract to reduce blood flow to the digestive system. This is part of the so-called **"fight-or-flight"** response of the sympathetic nervous system, allowing us to respond to dangerous situations.

The **adrenal cortex**, the outer portion of the adrenal gland, reacts to stress with the release of **corticosteroid** hormones that give a more long-term response. Corticosteroid hormones include both **glucocorticoids**, such as cortisol, and **mineralocorticoids**, such as aldosterone. All are steroid hormones. Glucocorticoids promote the synthesis of glucose from proteins and can suppress the immune system. The mineralocorticoids are involved in water and salt balance.

The **testes** of males and **ovaries** of females secrete sex hormones, which govern development and maintenance of sex organs as well as secondary sex characteristics. The testes make **androgens**, such as **testosterone**, and the ovaries make **estrogens**, such as **estradiol**, and **progestins**, such as **progesterone**. All are steroid hormones.

It should be evident that hormones play major roles both in animal development and in maintaining homeostasis in animals.

Topic Test 2: Examples of Hormonal Regulation

True/False

1. The adrenal gland contains both a medulla and a cortex.

2. Excess T_3 and T_4 can produce hypothyroidism.

3. Ovaries make estradiol and progesterone.

Multiple Choice

4. Which of the following hormones is most likely to have its receptors located inside of target cells?
   a. Epinephrine
   b. Insulin
   c. Glucagon
   d. Norepinephrine
   e. A corticosteroid

5. Which of the following regulates calcium levels in the blood?
   a. Calcitonin
   b. PTH
   c. Insulin
   d. Glucagon
   e. Both a and b

6. Which of the following is antagonistic to PTH?
   a. T_3
   b. Glucagon
   c. Insulin
   d. ADH
   e. Calcitonin

Short Answer

7. What do insulin and glucagon do to maintain homeostasis?

Topic Test 2: Answers

1. **True.** These are the two parts of the adrenal gland, and they secrete different hormones.

2. **False.** Hyperthyroidism would be produced.

3. **True.** These are sex hormones, made from cholesterol.

4. **e.** Corticosteroids are steroid hormones. Their receptors will be inside target cells, unlike peptide hormones (insulin, glucagon) or hormones made from amino acids (epinephrine, norepinephrine), which have receptors on the surface of target cells, as described in the first topic on hormone mechanisms.

5. **e.** Calcitonin reduces blood calcium levels, while PTH increases blood calcium levels.

6. **e.** Calcitonin and PTH are antagonistic hormones.

7. Insulin and glucagon are antagonistic hormones that regulate blood glucose concentration. Insulin stimulates an increased uptake of glucose by body cells, thereby reducing blood glucose levels. Glucagon increases breakdown of glycogen and release of glucose from the liver, thereby increasing blood glucose levels.

APPLICATION: DIABETES

Disorders in homeostasis for blood glucose are called **diabetes mellitus**. Typically, diabetes is caused either by low insulin levels or by target cells that are not adequately responsive to insulin. The result is high blood glucose concentrations.

There are two major types of diabetes. Type I is an early-onset version, also called juvenile diabetes, which is an **autoimmune** disorder. The immune system of a type I diabetic destroys the insulin-producing cells (beta cells) of the pancreas, leaving the individuals unable to generate adequate levels of insulin. Onset usually is sudden, during childhood. Insulin replacement is necessary for individuals with type I diabetes. Before the 1920s, when insulin was identified, most individuals with type I diabetes died.

Type II diabetes, also called adult-onset, or non-insulin-dependent, diabetes, is more commonly caused by reduced responsiveness to insulin in target cells. Overweight individuals who do not exercise are more prone to develop type II diabetes. Most diabetics are of this type, and many individuals manage to control their blood glucose levels by exercise and diet. Uncontrolled diabetes can lead to a number of other disorders and can reduce life span.

Chapter Test

True/False

1. Calcitonin is made in the thyroid gland.

2. T_4 is made in the thyroid gland.

3. Type II diabetes is an autoimmune disease.

Multiple Choice

4. Concerning insulin and glucagon,
   a. both increase blood glucose concentration.
   b. insulin increases blood glucose concentration, while glucagon decreases it.
   c. insulin decreases blood glucose concentration, while glucagon increases it.
   d. both decrease blood glucose concentration.
   e. neither one changes blood glucose concentration.

5. Which of the following is a second messenger that acts by releasing calcium from the endoplasmic reticulum?
   a. Cyclic AMP
   b. Testosterone
   c. IP_3
   d. Epinephrine

6. Steroid hormones
   a. bind to receptors on the surface of target cells, and act through second messengers.
   b. act directly on enzymes in the cytosol of target cells.
   c. are made from cholesterol.

d. bind to proteins in target cells and influence gene expression.

e. Both c and d are correct.

7. Which of the following serve as an antagonistic pair of hormones regulating blood glucose levels?

a. Insulin and glucagon

b. Insulin and cyclic AMP

c. Insulin and ACTH

d. Norepinephrine and epinephrine

Short Answer

8. What sequence of events occurs between the binding of epinephrine to a receptor and activation of protein kinases?

Chapter Test: Answers

1. **T**   2. **T**   3. **F**   4. **c**   5. **c**   6. **e**   7. **a**

8. Epinephrine binds to a receptor in the membrane of the target cell. The receptor-hormone complex activates G proteins, which split GTP and activate adenylyl cyclase. The adenylyl cyclase generates cyclic AMP from ATP. The cyclic AMP activates protein kinases.

Check Your Performance:

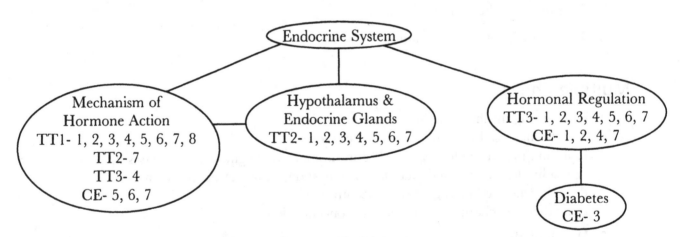

Key: TT = Topic Test; CE = Chapter Exam. Numbers indicate exam questions. Some questions are listed more than once if they refer to more than one topic.

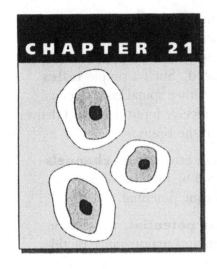

CHAPTER 21

Neurons: Electrical and Synaptic Events

Neuronal and glial cells are the two major kinds of cells in the nervous system. Neurons, or nerve cells, are the elements of Nature's computers. Glial cells play a supporting and insulating role in the mature nervous system, and also play roles during development and regeneration of nervous system tissue.

Neurons typically contain a **cell body** (with the nucleus of the cell), an elongated **axon**, and a set of **dendrites** that, with the cell body, receive signals. These signals are electrical in nature within neurons, which propagate **action potentials (nerve impulses)** along their axons from one end to the other. Understanding the electrical signals in neurons requires knowledge about electrical currents (the movement of electrical charges, which are ions such as sodium and potassium), **electrical potentials** (voltage differences across membranes that can be produced by the movement of the ions), and **conductance** (the reciprocal of **resistance**), which increases as the number of open channels in the membrane increases.

We examine the nature of electrical signals in neurons and then consider the mechanism of chemical **synaptic transmission** that neurons use to communicate with each other and with other cells.

ESSENTIAL BACKGROUND

- Current, potential, and conductance
- Permeability (Chapter 4)
- Concentration gradients (Chapter 4)
- Na-K ATPase (Chapter 4)

TOPIC 1: NEURONS, CIRCUITS, AND RESTING POTENTIALS

KEY POINTS

✓ *What is a typical reflex circuit in the spinal cord?*

✓ *What is the resting potential and what is required for its generation and maintenance?*

✓ *What are equilibrium potentials?*

Functionally, the **nervous system** is located between **sensory inputs** and **motor outputs**. Sensory inputs give the nervous system information about the body and the environment. The motor outputs allow the nervous system to control behavior through muscle contractions and glands.

In the nervous system, neurons are linked together into **circuits**. Simpler circuits in vertebrates are located in the spinal cord and control reflexes. The simplest such circuit has a sensory neuron receiving input from a sensory receptor (such as a stretch receptor in a muscle), and a motor neuron (which might cause muscle contraction when stimulated). Such a simple **reflex circuit** causes the **knee-jerk (tendon-tap) reflex**, for example. Other spinal cord circuits involve **interneurons** (interneurons are simply neurons that both receive input from and deliver output to other neurons). More complex neural circuits are found in the brain.

The key to understanding electrical events in neurons is to understand the various **channels** that are present in nerve cells. Electrical events in neurons are caused by the flow of ions through such channels. Those currents cause changes in the membrane potential.

Neurons maintain a potential (voltage) difference, called the **resting potential**, across their outer membranes. Three major factors contribute to the production and maintenance of this resting potential:

1. A large number of open **potassium (K⁺) channels** in the membrane. The resulting high resting permeability to potassium allows potassium ions to flux easily across the membrane—in other words, conductance for potassium is high.

2. A concentration gradient for potassium ions. There is a higher concentration of potassium ions inside the neuron than outside.

3. The Na-K ATPase. This sodium-potassium pump maintains the concentration gradients for potassium and sodium across the membrane.

We can see how the resting potential develops if we start with a concentration gradient for potassium and no potential across the membrane. Potassium ions will move down their concentration gradient by diffusion across the membrane, through the open potassium channels, with more of the ions flowing from inside to outside. That net flux of positively charged potassium ions will cause an electrical potential to build, with the inside becoming negative relative to the outside. As the negative charge builds on the inside, and the positive charge builds on the outside of the neuron, the membrane potential will begin to oppose the flow of more positively charged potassium ions from inside to outside, and the flow will slow down.

The concentration gradient for potassium can produce a membrane electrical potential of about −75 millivolts (mV; inside negative relative to outside) at equilibrium (when the driving force of the concentration gradient is just offset by the voltage gradient). This is called the **equilibrium potential** (Chapter 4) for potassium, the potential that just offsets the concentration gradient, and causes the fluxes of potassium to be equal into and out of the cell.

The resting potential of the cell is not quite as negative as the potassium equilibrium potential because there is a slight permeability to sodium ions (Na⁺), which have a concentration gradient in the opposite direction to that of potassium, because there is more sodium outside than inside. Sodium ions are at equilibrium when the potential across the cell is about +55 mV. A typical neuron's resting potential of −70 mV is much closer to the potassium equilibrium potential than to sodium's because potassium is much more permeable than sodium (more resting potassium channels are open).

There is a constant, slight leakage of sodium and potassium across the membrane as each tries to drive the potential toward their equilibrium potentials. That leakage is pumped back, sodium ions out and potassium ions in, by the **Na-K ATPase** (Chapter 4). This molecular pump, driven by hydrolysis of ATP, is responsible for maintaining the concentration gradients of the two ions, which in turn maintain the resting potential. All of the values of potentials, such as those for the resting potential and equilibrium potentials, given in this chapter should be considered estimations. They can vary significantly among neurons.

Topic Test 1: Neurons and Resting Potentials

True/False

1. Neurons use chemical transmitters to send signals along their axons.

2. An equilibrium potential for an ion is the potential across the membrane of the neuron that is able to offset the concentration gradient for the ion, resulting in no net flux across the membrane for that ion.

3. The resting potential of a neuron is very close to the equilibrium potential for sodium ions.

Multiple Choice

4. If the membrane potential (voltage) is significantly different from the equilibrium potential for an ion,
   a. the ion will flow across the membrane if the conductance for the ion is zero.
   b. a driving force will be produced, and current will flow if the ion is permeable across the membrane.
   c. no current can flow.
   d. no force will be generated because there only is a net flow of ions when the equilibrium potential is equal to the membrane potential.
   e. both a and c are correct.

5. The resting potential across a neuron
   a. involves membrane channels that allow potassium ions (K^+) to cross the membrane.
   b. requires a higher concentration of K^+ outside of the cell than inside.
   c. requires the Na-K pump to maintain it.
   d. is zero, since everything is at rest.
   e. Both a and c are correct.

Short Answer

6. Describe the movement of sodium and potassium ions across a neuron's membrane during the resting potential. What does the Na-K ATPase do to this flux of ions?

Topic Test 1: Answers

1. **False.** Neurons transmit electrical signals along their axons. They use chemical transmission to send signals between cells.

2. **True.** The equilibrium potential specifies the membrane potential at which the net flux of an ion is zero. At the equilibrium potential, the electrical and chemical gradients generate forces that are equal and opposite.

3. **False.** The resting potential is much closer to the equilibrium potential for potassium than for sodium.

4. **b.** Current will flow if the conductance is not zero, that is, if there are channels for the ion that are open. The current will flow in a direction that will bring the membrane potential toward the equilibrium potential for that ion.

5. **e.** The large number of open K^+ channels maintains the membrane potential close to the potassium equilibrium potential. The Na-K pump is necessary to maintain the concentration gradients for potassium that, in turn, determines the equilibrium potentials for Na and K.

6. There is a slight "leakage" of sodium ions and potassium ions during the resting potential. Sodium diffuses from outside, where it is in higher concentration, to inside. Potassium diffuses in the opposite direction, since it is more concentrated inside the neuron. The Na-K pump maintains the concentration gradients for the two ions by pumping Na^+ out of the cell and K^+ back into the cell.

TOPIC 2: THE ACTION POTENTIAL

KEY POINTS

✓ *What is an action potential, and what causes it to occur?*

✓ *How does the action potential propagate down an axon?*

✓ *What role do glial cells play in action potential propagation?*

Nerve cells send signals from one part of themselves to another. These signals are all-or-nothing electrical events called **action potentials** or **nerve impulses**. Typically these action potentials will **propagate**, or move, along the axon, from the nerve cell body where they first are generated to the synaptic terminals at the end of the axon. The action potential consists of a series of quick changes in the potential of the membrane (**Figure 21.1**).

Sets of special sodium and potassium channels in the membrane underlie the action potential. These channels are **voltage sensitive**, or **voltage gated**; that is, they open as a result of the membrane potential reaching a certain "threshold" value. This threshold voltage is about −60 mV, or about a 10-mV depolarization (from negative toward zero) from the resting potential.

A **positive feedback loop** both triggers the action potential and allows it to propagate (spread down the axon). This loop consists of 1) voltage depolarization, causing 2) the opening of **voltage-gated sodium channels**, causing 3) the net influx of sodium ions into the cell, causing 4) a further depolarization of the cell. As you can see, step 4 returns us back to step 1, thus completing the positive feedback loop. The opening of the sodium channels allows the movement of sodium ions into the cell, which changes the potential across the membrane. That potential change is the start of the action potential.

During the action potential, the membrane voltage depolarizes from the resting potential of about −70 mV, and actually climbs to about +25 mV. This happens because as sodium becomes more permeable than potassium, the potential is driven toward the sodium equilibrium potential of +55 mV. The potential does not reach above about +25 mV for two reasons. First, a few tenths of a millisecond after opening, the sodium channels inactivate; that is, they no longer allow sodium ions to pass through the membrane. Second, **voltage-gated potassium channels**

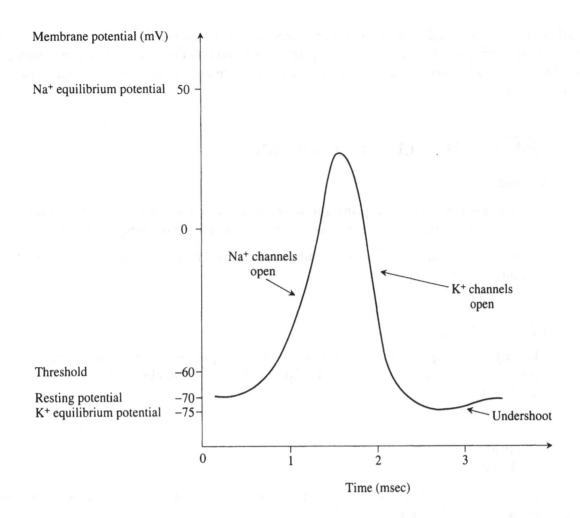

Figure 21.1 The action potential. The action potential is a voltage change that occurs over time in a neuron. It is produced by the opening of voltage-gated sodium and potassium channels.

begin to open. That opening is delayed and occurs at about the same time as the sodium channels are inactivating. These voltage-gated potassium channels, which differ from those underlying the resting potential, increase potassium flow out of the neuron and drive the membrane potential rapidly back toward the resting potential. In fact, potassium becomes so much more permeable than sodium that the membrane potential exhibits an "undershoot," more negative than the resting potential, as the membrane potential gets very close to the potassium equilibrium potential. As these voltage-gated potassium channels close, the membrane potential returns to the resting potential. This completes one action potential. The sodium channels are reactivated during the membrane potential repolarization, thus preparing the cell for another action potential.

The entire action potential, from beginning to end, lasts about 1 millisecond. Some nerve cells can "fire" more than 500 action potentials in a second (500 hertz). Each action potential looks much like the last, and thus these are called **all-or-nothing** potentials. Once initiated, the action potential maintains its amplitude as it propagates down the axon because of the depolarizing currents that spread ahead of the action potential and bring the membrane potential to threshold farther down the axon. Thus, the action potential is **self-propagating**.

Glial cells form **myelin sheaths** around some axons. These multiple wraps of membrane increase the efficiency and the velocity of propagation (conduction velocity) of the action potential by insulating the axon, that is, by increasing resistance and decreasing capacitance of the axonal membrane.

Topic Test 2: The Action Potential

True/False

1. The action potential is self-propagating because it can regenerate itself as it moves down an axon, and thus maintains the same voltage amplitude over long distances.

2. During the action potential, the membrane potential shifts from threshold toward the sodium equilibrium potential.

Multiple Choice

3. During the rising (depolarizing) phase of an action potential, the ion that is most permeable (highest conductance) through the nerve cell membrane is
   a. sodium.
   b. potassium.
   c. chloride.
   d. calcium.
   e. phosphate.

4. During the undershoot of an action potential, the ion that is most permeable through the nerve cell membrane is
   a. sodium.
   b. potassium.
   c. chloride.
   d. calcium.
   e. phosphate.

Short Answer

5. Describe the positive feedback loop that underlies the initiation and propagation of nerve impulses.

Topic Test 2: Answers

1. **True.** The existing ion gradients represent the stored, potential energy that is used for the self-propagation. The potential is all-or-nothing because of this self-regeneration.

2. **True.** As the voltage-gated sodium channels open, the potential moves in the direction of the sodium equilibrium potential, but does not reach it for the two reasons given in the text.

3. **a.** Sodium channels have opened, and increase the conductance of sodium ions above that of all other ions.

4. **b.** Potassium ions are most permeable because so many potassium channels are open at this time.

5. A decrease in the magnitude of the resting potential (depolarization) causes voltage-sensitive sodium channels to open. The open channels allow Na^+ to diffuse into the cell. The positively charged sodium ions further depolarize the cell, opening more channels, which allow more Na^+ to flow, and so on.

TOPIC 3: RECEPTOR POTENTIALS AND SYNAPTIC POTENTIALS

KEY POINTS

✓ *How do action potentials get generated in neurons?*

✓ *What are passive potentials and how do they spread?*

✓ *What are EPSPs and IPSPs and how do they arise?*

Unlike the action potential, other membrane potentials spread **passively**. These potentials are not all-or-nothing, but instead diminish in amplitude as they propagate from their sites of initiation. Two important classes of such passive potentials in neurons are receptor potentials and synaptic potentials.

Receptor potentials, also known as **generator potentials**, are produced by sensory transducers, which are specialized receptor cells, or parts of such cells, that respond to environmental input by opening ion channels in membranes. The resulting flux of ions through the open channels produces the receptor potential. There can be a complex set of steps that link the sensory stimulus to the receptor potential. A strong enough receptor potential can reach threshold and produce an action potential.

The other kind of passively spread potential is a **synaptic potential**, which occurs in a postsynaptic neuron as a result of the binding of chemical transmitters to specialized receptors on the membrane of the cell. Synapses are located on the dendrites and cell bodies of neurons. We consider the process of synaptic transmission in detail in the next section. The chemical transmitters are released from the sender (presynaptic) neuron and diffuse to the receiver (postsynaptic) cell. The result of the process of synaptic transmission is the generation of synaptic potentials that influence the production of action potentials in the postsynaptic cell.

There are two kinds of synaptic potentials. Some are excitatory; that is, they tend to depolarize the neuron toward threshold. These are called **excitatory postsynaptic potentials (EPSPs)**. Others are inhibitory and tend to hold the membrane potential below threshold. These are called **inhibitory postsynaptic potentials (IPSPs)**. EPSPs can be generated by opening channels that allow the passage of both sodium and potassium ions. This kind of ion channel is also found at nerve-muscle synapses where it is linked to receptors for the chemical acetylcholine. Inhibitory receptors are linked to channels that typically increase permeability to either chloride ions (Cl^-) or potassium ions. Because of their concentrations inside and outside of the cell, either one of these ions, when permeable, will tend to "lock" the membrane potential below threshold, thereby inhibiting production of an action potential.

Typical neurons are being bombarded with IPSPs and EPSPs constantly. In a typical neuron, the firing of action potentials occurs when the part of the cell body leading to the axon, a zone called the **axon hillock**, reaches the threshold potential. A **spatial** and **temporal summation** of the postsynaptic potentials is constantly occurring in the neuron. Spatial summation occurs when the passively spreading currents from two or more synapses interact. Temporal summation occurs when a given synapse fires repeatedly, allowing for a buildup of the postsynaptic potential beyond that of a single input. Both spatial and temporal summation occur continuously in nerve cells and can produce trains of action potentials or can suppress firing entirely. The firing pattern and firing frequency of neurons carry meaning for the brain and govern behavior.

Topic Test 3: Receptor Potentials and Synaptic Potentials

True/False

1. Synaptic potentials are all-or-nothing, self-propagating potentials.

2. Spatial and temporal summation refers to the summing of IPSPs and EPSPs.

3. Receptor potentials are related to the transduction of sensory stimuli.

Multiple Choice

4. Receptor potentials and EPSPs
   a. are examples of active potentials, spread by positive feedback.
   b. are examples of passive potentials whose voltages decrease as they spread along a nerve cell.
   c. involve voltage-gated sodium channels.
   d. can produce action potentials in a nerve cell if they allow the cell to reach threshold.
   e. include both b and d.

5. Concerning summation of postsynaptic potentials,
   a. IPSPs tend to inhibit or prevent EPSPs from being able to cause neurons to fire action potentials.
   b. EPSPs and IPSPs can occur at the same time in a neuron.
   c. two EPSPs occurring at slightly different times at the same synapse can sum to generate an action potential.
   d. two EPSPs occurring at different synapses on the same neuron can sum to generate an action potential.
   e. all of the above are correct.

Short Answer

6. Describe how passive potentials can lead to action potentials.

7. Compare active and passive propagation of electrical signals in neurons.

Topic Test 3: Answers

1. **False.** Synaptic potentials propagate passively.

2. **True.** Such synaptic potentials sum over time and space in a neuron.

3. **True.** Receptor potentials are caused by sensory stimulation and initiate the process of sensory input to the nervous system.

4. **e.** These are passive potentials that can produce action potentials if they are strong enough.

5. **e.** All of the statements are correct and can be used to check your understanding of summation. (From a test-taking perspective, also note that when one answer is "all of the above," you only need to be certain of two of the answers to know that this is the right answer—or that the question will be thrown out because there is more than one right answer!)

6. Receptor potentials and EPSPs can depolarize the membrane from the resting potential to threshold. At that point enough voltage-sensitive sodium channels have opened to sustain the development of an action potential by positive feedback.

7. Active propagation occurs with the action potential. Active propagation is reinforced as it passes along the neuron, and so maintains a constant amplitude. Passive potentials are not reinforced and consequently diminish in amplitude with distance.

TOPIC 4: SYNAPTIC TRANSMISSION

KEY POINTS

✓ *What sequence of events occurs during synaptic transmission?*

✓ *What ion is important in triggering the release of synaptic transmitter from the presynaptic neuron?*

✓ *Where is chemical transmitter substance stored in the presynaptic terminal?*

✓ *What are some of the chemicals that serve as synaptic transmitters?*

Synaptic transmission is the process by which a presynaptic neuron signals a postsynaptic cell. Synaptic transmission takes about 1 millisecond to complete and consists of the following sequence of events:

- Action potential occurs in the presynaptic neuron.
- The presynaptic terminal depolarizes.
- The voltage-gated calcium (Ca^{2+}) channels open.
- Ca^{2+} enters the presynaptic terminal.
- Synaptic vesicles release transmitter from the presynaptic terminal into the synaptic cleft.
- Chemical transmitter diffuses across the cleft.
- Transmitter binds to receptors on postsynaptic cell membrane.
- Ion channels linked to the receptors open.
- EPSP or IPSP occurs.

The **transmitter** substance binds reversibly to the receptor and when released, depending on the transmitter, either is degraded by enzymes in the cleft or is taken back up by active transport into the presynaptic cell for reuse.

EPSPs and IPSPs last for 1 or a few milliseconds and typically change membrane potential by 1 or a few millivolts. The magnitude of the voltage change declines as the potentials spread from the synapse along the surface of the postsynaptic cell.

Common transmitter substances include the following:

1. **Acetylcholine** is found in the central nervous system and some of the cells that make it may die selectively in persons with Alzheimer's disease. It also is the transmitter at neuromuscular junctions.

2. **Epinephrine**, **norepinephrine**, and **dopamine** are chemically related substances that play a role in mood and emotion in humans. Dopamine also is found in certain motor pathways in the brain, and the cells that produce it selectively die in persons with Parkinson's disease. Dopamine-containing brain pathways also play a role in schizophrenia.

3. **Serotonin** also plays a role in mood and is involved in sleep. The antidepressant drug Prozac may act by blocking serotonin reuptake at synapses, thereby increasing its influence on postsynaptic cells.

4. **GABA (gamma-aminobutyric acid)** is an inhibitory neurotransmitter in humans. If there is insufficient GABA in the brain, seizures can result.

There are many other transmitter substances. Some short proteins, called **neuropeptides**, also are released at some synapses during synaptic transmission and play a role in the modulation of neuronal activity in the brain. Some of these are involved in pain perception.

Topic Test 4: Synaptic Transmission

True/False

1. The release of synaptic transmitter is triggered by calcium ions entering the synapse from outside of the neuron.

2. In Parkinson's disease, the level of acetylcholine in the brain decreases more than does the level of other transmitter substances.

Multiple Choice

3. Synaptic transmission in neurons
   a. involves the release of synaptic transmitter from vesicles.
   b. requires Ca^{2+}.
   c. requires receptors on the postsynaptic membrane.
   d. produces an IPSP or EPSP in the postsynaptic cell.
   e. includes all of the above.

4. Signal propagation from one neuron to another in the brain most often involves
   a. an electrical signal passing from the first to the second neuron.
   b. a chemical signal passing from the sending to the receiving neuron.
   c. sodium ions leaving one cell and entering the other.
   d. potassium ions leaving one cell and entering the other.
   e. both a and c.

Short Answer

5. In synaptic transmission, what happens after calcium ions enter the presynaptic terminal?

Topic Test 4: Answers

1. **True.** These calcium ions enter through channels that open when the presynaptic cell is depolarized by the action potential.

2. **False.** During Parkinson's disease it is certain dopamine-containing neurons in the brain that die. Physicians treat Parkinson's disease with dopa, a temporarily helpful compound that can enter the brain and be used to make dopamine by any remaining dopamine-containing neurons.

3. **e.** All of the statements are correct, as can be seen by examining the list in text of the events involved in synaptic transmission.

4. **b.** The details of the process of a chemical transmitter acting between neurons are provided in this section. While sodium and potassium ions do enter and leave cells, these movements are not what carry the signal from the presynaptic to the postsynaptic cell. Rather, the chemical transmitter allows for ion movements in one cell to trigger ion movements in another cell. A few neurons do communicate by direct electrical contact through gap junctions (Chapter 5), but these are a small minority.

5. After Ca^{2+} ions enter, they trigger the fusion of synaptic vesicles, which are in the presynaptic terminal, with the outer membrane of the presynaptic neuron. This allows chemical transmitter, located in the vesicles, to be released from the cell.

APPLICATION: BIOLOGY OF MIND

Our current knowledge of how neurons work leaves us with perhaps the most puzzling unexplained phenomena in the universe: mind and consciousness. Many neuroscientists today accept that mental abilities, such as consciousness, thinking, understanding, perceiving, learning, and being aware, are the result of brain activity, or basically, neuronal activity. How do personal experiences arise from the action potentials or other activities of billions of individual neurons? What holds things together, giving us a seemingly unified view of the world and ourselves? How does mind arise from matter?

These questions have not been answered, and some have questioned whether they are the right questions to ask. We have some idea of how memory and learning work and have maps of the brain indicating the various regions where particular activities, such as visual information processing, are carried out. We know something of the language processing centers in the brain, and of the differences in functioning between the left and right cerebral hemispheres. Modern brain imaging techniques, such as functional magnetic resonance imaging (fMRI) and positron emission tomography (PET), have given us views of "where the action is" in the brain as we carry out various mental tasks. None of this has given us ultimate insight into what appears to be a great gap between neural correlates of mental activity and an understanding of why we have experiences. There is a difference between what our brains accomplish and what computers can do, at least up to now—we experience, we understand, and we grasp meanings.

Until our experiences, and our mental abilities more generally, can be integrated into the science of the physical universe, our world view will remain incomplete. If we can gain such an integration, the resulting insights may advance our understanding of human nature as much as our knowledge of DNA has advanced the science of genetics.

Chapter Test

True/False

1. Receptor potentials and EPSPs can produce action potentials.

2. Acetylcholine and dopamine are synaptic transmitters.

3. Some synaptic transmitters are destroyed in the synaptic cleft.

Multiple Choice

4. Which of the following is NOT involved in generating or maintaining the resting potential?
   a. Open potassium channels
   b. Concentration of potassium across the membrane
   c. The Na-K ATPase
   d. The potassium channels involved in the action potential

5. What role does calcium play during synaptic transmission?
   a. It is released from the endoplasmic reticulum and triggers the action potential.
   b. It is released from endoplasmic reticulum by a second messenger.
   c. It enters the neuron and triggers release of synaptic transmitter by vesicle fusion.
   d. It is a synaptic transmitter.

Short Answer

*6. What would happen to the resting potential of a neuron if one were to add potassium chloride to the fluid outside of the neuron?

Chapter Test: Answers

1. **T** 2. **T** 3. **T** 4. **d** 5. **c**

6. The addition of potassium ions would reduce the concentration gradient for potassium. The resting potential is largely determined by that gradient, and the resulting potassium equilibrium potential. That potential would be less negative, and so would the resting potential. The final value (−60, −50, or whatever) depends on how much potassium is added.

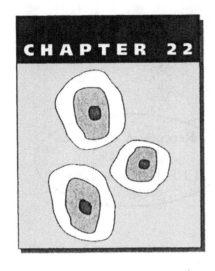

CHAPTER 22

Immune System and Body Defenses

Animals have a number of natural defenses against disease. We defend ourselves against intruders as well as abnormal body cells. This chapter stresses body defenses found in mammals; similar defenses are found among other vertebrates. Invertebrates have different systems.

There are several levels of nonspecific defenses plus specific defense mechanisms. Sometimes the term **immune system** is used to refer only to the specific defense mechanisms, which are targeted toward specific entities or structures. The nonspecific defenses are more general in their targeting.

The lymph system was mentioned in Chapter 19. It is part of the immune system, and so are bone marrow, thymus, spleen, and certain constituents of blood.

ESSENTIAL BACKGROUND

- Type I diabetes (Chapter 20)

TOPIC 1: THE BODY'S DEFENSE MECHANISMS

KEY POINTS

✓ *What do our external defenses include?*

✓ *What are phagocytic cells?*

✓ *What is the inflammatory response?*

✓ *What are the specific defense mechanisms?*

Our skin and mucous membranes offer the first line of defense against infections. Skin offers a penetration barrier, and secretions from mucous membranes can protect by washing away intruders or by inhibiting, trapping, or killing them through acidic pH, enzymes, and viscous fluids.

Several kinds of white blood cells (**leukocytes**) are general phagocytic cells. **Neutrophils** make up about two-thirds of all white blood cells, and can ingest invading particles by phagocytosis, a form of endocytosis (see Chapter 4). Neutrophils tend to self-destruct as they destroy foreign invaders. They have a half-life of only a few days, so they are constantly being made by the body. **Monocytes**, comprising about 5% of white blood cells, also ingest foreign invaders and kill with digestive enzymes and reactive forms of oxygen. Monocytes can mature into larger,

Check Your Performance:

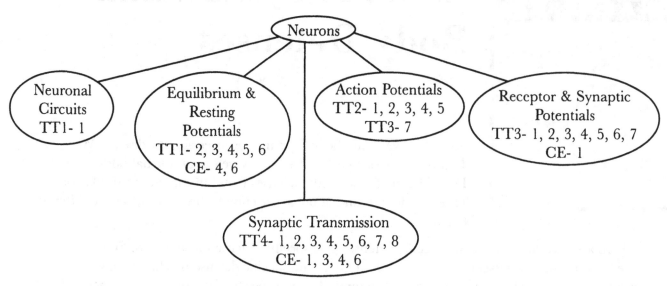

Key: TT = Topic Test; CE = Chapter Exam. Numbers indicate exam questions. Some questions are listed more than once if they refer to more than one topic.

longer-lived phagocytes, called **macrophages**, which also can become **antigen-presenting cells**, as we discover in the next section. **Eosinophils** are present in lower abundance. They attach to larger parasites, such as worms, and release destructive enzymes from granules by exocytosis (Chapter 4). **Natural killer cells**, another form of leukocyte, destroy infected and cancerous cells in the body.

In addition, antimicrobial proteins, including the **interferons**, can be produced by virus-infected cells and help other cells resist infection.

Complement is a group of more than 20 proteins that can act together to lyse (break open) invading microbes. The proteins act in sequence to create a hole in the foreign cell. Complement also can signal phagocytes by binding to invading microbes or by triggering the release of histamine, thereby activating the inflammatory response.

The **inflammatory response** most often is triggered by tissue damage. For instance, the skin might be cut. The injury triggers the release of histamine from mast cells and other cells. Histamine is a local regulator (Chapter 20) that dilates nearby blood vessels and increases their permeability. This allows fluid, blood clotting elements, and phagocytes to migrate to the site of injury, causing swelling and redness. The phagocytes consume bacteria and other microorganisms, thereby protecting against infection while the wound heals.

The functions of the specific defense mechanisms are to recognize, eliminate, and remember foreign invaders and abnormal body cells. There are two related systems. One produces protein **antibodies** and is called **humoral**. The other generates **cell-mediated** immunity. We examine these in detail in the next section.

Some important concepts related to specific defense mechanisms include the following:

1. **Specificity**—particular microorganisms or molecules are recognized and eliminated. Their shapes, called **antigenic determinants**, trigger the responses. These determinants are parts of molecules found on **antigens**, which might be whole molecules or multimolecular viruses, and so on.

2. **Diversity**—millions of different kinds of invaders or altered cells can be responded to.

3. **Self and non-self distinction**—there is an ability to distinguish between the shapes of one's own molecules and those of invading or abnormal entities.

4. **Immunological memory**—immune systems react more quickly to repeat contact with a particular invader.

Topic Test 1: The Body's Defense Mechanisms

True/False

1. Monocytes are the most abundant type of white blood cells.

2. Monocytes can be macrophages, which can be antigen-presenting cells.

3. Complement consists of proteins that can destroy cells.

4. The immune system has a memory.

5. Antigenic determinants are the keys to the specificity of the immune system.

Multiple Choice

6. The two systems of specific defense mechanisms are
   a. specificity and diversity.
   b. inflammatory response and complement.
   c. antigens and antigenic determinants.
   d. humoral and cell mediated.
   e. monocytes and phagocytes.

Topic Test 1: Answers

1. **False.** Neutrophils are the most abundant.

2. **True.** All three are the same cell in different stages. Macrophages, when they digest foreign entities, are able to present parts of the foreign entity on their surfaces, and thereby become antigen-presenting cells, as we will see later.

3. **True.** Complement proteins can cause cells to lyse.

4. **True.** The immune system is able to recognize and respond more rapidly to re-exposure to an entity.

5. **True.** The immune system recognizes particular antigenic determinants, which are parts of antigens. We examine how in the next section.

6. **d.** Humoral immunity involves the production of antibodies, and cell-mediated immunity involves the destruction of infected and abnormal body cells.

TOPIC 2: B CELLS, T CELLS, AND THE IMMUNE RESPONSE

KEY POINTS

✓ *What are the two major kinds of lymphocytes?*

✓ *What are antibodies, how are they generated, and what do they do?*

✓ *What are memory cells?*

✓ *How does cell-mediated immunity work?*

Among the leukocytes (white blood cells) are **lymphocytes**. There are two major kinds: **B cells**, or **B lymphocytes**, are responsible for humoral immunity, that is, antibody production. Antibodies are proteins that protect against free viruses, bacteria, and toxins. **T cells**, or **T lymphocytes**, are responsible for cell-mediated immunity, and also play a role in the stimulation of B cells. Cell-mediated immunity is active against cells, be they bacteria, fungi, protozoans, worms, body cells infected by viruses, cancer cells, or transplanted cells.

All blood cells originate in bone marrow. B cells mature in bone; T cells migrate to the thymus to mature. T and B cells have antigen receptors on their membranes, giving them specificity and diversity. B-cell receptors are antibodies. Antigen binding to the receptors on B and T cells can lead to cell multiplication of the specific B and T cells recognizing the antigen, and the B and T cells can then become active effector cells. However, there are complexities to this antigen binding.

When activated, B cells give rise both to **plasma cells**, which generate large numbers of antibodies, and to **memory cells**, which protect against future infections by the same entities.

T cells include **cytotoxic T (T$_C$) cells** (also known as killer T cells), which destroy infected and cancer cells; **helper T (T$_H$) cells**, which stimulate humoral and cell-mediated immunity; and **memory T cells**, which protect against future infections.

A **primary immune response** requires 5 to 10 days from the time of exposure to maximum production of effector cells. A **secondary immune response** is based on memory cells, and requires 3 to 5 days. It can occur weeks or years after the primary immune response has produced memory cells.

Antigens typically are proteins, polysaccharides, or other novel shapes to the individual. Often these are on the surfaces of invaders. **Antibodies** are **immunoglobulins**. Each antibody has variable and constant regions. The constant regions are shared among all antibodies of a given type. In mammals, there are five types of constant regions, corresponding to the five different classes of immunoglobulins. The variable region is specific for a particular antigenic determinant, and binds to it. Each antibody has at least two identical binding sites. In brief, the antibodies bind to antigenic determinants on antigens. Interesting genetic modifications give rise to the many variable regions found in antibodies.

Individuals have a large variety of different kinds of B cells, and each makes only one kind of antibody. Activation of a B cell requires binding of antigens to the receptors on the specific B cell. A given type of antigen can activate only a small number of B cells that recognize one or a few antigenic determinants on the antigen. Activated B cells produce a clone of plasma cells, each of which generates about 2,000 copies of one specific antibody each second for their 4- to 5-day lifetime. The activated B cells also produce a clone of memory cells, which can give rise to the secondary immune response. Usually, T$_H$ cells as well as antigen are required to stimulate the B cells. That process is described later.

Once produced, antibodies act in several different ways. By binding to virus antigens, they can block the ability of virus to bind to cells. By binding to a toxin, an antibody can block the action of that toxin. In either case, phagocytic cells eventually will dispose of the antigen-antibody complex. Because they have multiple binding sites, antibodies also can produce clumps of antigen-antibody complexes. Phagocytic cells then can engulf the cluster. Finally, binding of antibodies to the surface of an invading cell can activate complement, which, as described earlier, can result in the bursting of the cell.

To learn how T cells work, we first must learn about the **major histocompatibility complex (MHC)**. This complex comprises the set of cell surface glycoproteins found on most body cells. They are made from at least 20 genes and there are at least 50 alleles for each gene in the human population. This means that the MHC glycoproteins are very specific and individualistic. **Class I** MHC proteins are found on all cells in the body with nuclei. **Class II** MHC proteins are found on macrophages and B lymphocytes.

Macrophages engulf pathogens, digest them, and display antigens from the pathogen on their surface, cradled by class II MHC molecules. Such macrophages are called **antigen-presenting cells**. T$_H$ cells are activated when their receptors recognize the combination of specific antigen and surrounding class II MHC proteins on antigen-presenting cells. As a result of the contact, both the macrophage and the T$_H$ cell release protein **cytokines**, including **interleukins**, which stimulate production of a clone of T cells. The newly produced T cells are capable of interacting with B cells.

B cells also present antigen on their surfaces. An activated T_H cell will bind only if it recognizes the B cell's antigen and class II MHC. The T_H cell then releases interleukins that stimulate the B cell to produce plasma cells. For most antigens, such T-cell activation is required along with antigen presence to stimulate B-cell production. A few antigens are able to induce plasma and memory cell production in B cells without help from T cells.

T_H cells also activate T_C cells, which also can be activated by coming in contact with infected or cancerous cells that are displaying antigens recognized by the T_C cell along with class I MHC. Notice that the T_C cells are recognizing "self" in the form of class I MHC complexed with "non-self" antigens on the surface of the cells, which might, for example, be infected with a virus. Non-self antigens also can be found in cancer cells because mutations in such cells (Chapter 11) can produce altered proteins that are recognized as "non-self" by the T cells. This induces T_C-cell proliferation. Activated T_C cells then can bind to infected or abnormal cells, recognizing the combination of specific antigen and class I MHC, and release **perforin**, a protein that can lyse (break open) the cell. In this way, virus-infected cells can be destroyed before they can produce more virus, and some cancer cells can be destroyed as well.

 For information on **blood groups**, **allergies**, **transplants**, and **autoimmune diseases**, see the web page.

Topic Test 2: B Cells, T Cells, and the Immune Response

True/False

1. Cytokines, including interleukins, are released by T_H cells.

2. Each antibody contains a single binding site for an antigen.

3. The secondary immune response occurs more rapidly after antigen exposure than the primary response.

Multiple Choice

4. In the immune system, some T lymphocytes
   a. synthesize and release antibodies, each T cell releasing a specific antibody.
   b. synthesize and release antibodies, each T cell releasing a variety of different kinds of antibodies.
   c. can attack and destroy infected cells by lysis.
   d. lack specificity and attack any cell that is nearby.

5. Complement
   a. can be activated by antibodies bound to a foreign cell.
   b. consists of a set of proteins.
   c. can cause cells to break apart.
   d. includes both b and c.
   e. includes all of the above.

6. Major histocompatibility complex proteins
   a. are cell surface glycoproteins that differ in most humans.
   b. include class I on the surface of most cells.
   c. include class II on the surface of macrophages and B and T cells.

d. are recognized by T cells.

e. are all of the above.

7. Memory cells in the immune system

a. are only made when there is a second exposure to an antigen.

b. are made during a first exposure to an antigen and protect the body against future exposures to the same antigen.

c. are only T_C cells.

d. are both b and c.

8. Concerning the immune system,

a. plasma cells make antibodies.

b. T_C cells can destroy virus-infected cells.

c. activated T_H cells stimulate B cells.

d. complement can trigger cell lysis.

e. all of the above are correct.

Topic Test 2: Answers

1. **True.** These proteins stimulate B cells and other T cells.

2. **False.** Antibodies contain two or more such sites, depending on the kind of antibody.

3. **True.** It takes less time to respond after the first exposure to an antigen.

4. **c.** These are the cytotoxic T cells.

5. **e.** These answers form a foundation of information about complement.

6. **e.** These answers form a foundation of information about MHC proteins.

7. **b.** The memory cells are the basis of the secondary immune response.

8. **e.** All are correct.

APPLICATION: IMMUNODEFICIENCY

A variety of conditions can cause immunodeficiency, which increases vulnerability to infection or cancer. Some cancers depress the immune system, which actually can prevent the cancer from being destroyed by the immune system. Physical and emotional stress also can suppress the immune system. Because of the increased physical stress, athletes who train very hard can actually increase their risk of infection.

A recently growing risk is **acquired immunodeficiency syndrome (AIDS)**, which is caused by human immunodeficiency virus (HIV). The mortality rate for those with the virus approaches 100%, although recent combinations of drug therapies can extend life expectancy, at least in the short run. HIV infects T_H cells, thereby damaging the immune system. The body produces more T_H cells, but the virus, which can mutate quickly, manages to continue to infect the cells. A steady state develops and can last years, with new T cells being produced and then infected. Eventually the body cannot continue to produce enough T cells, and the number of T_H cells declines to such a low level that cell-mediated immunity collapses, and individuals succumb to one or more of a variety of possible infections. Given the presence of AIDS, participating in unprotected sex, except for making babies, is not recommended.

Chapter Test

True/False

1. T cells are the same as T lymphocytes.

2. Lymphocytes are types of white blood cells.

Multiple Choice

3. B cells in the immune system
   a. make plasma cells.
   b. are stimulated by T_H cells.
   c. are stimulated by antigens.
   d. have antibodies on their surface.
   e. include all of the above.

4. Immunological memory refers to
   a. the ability of our immune systems to respond to the same antigen that our parents had contact with years earlier.
   b. our immune system's ability to react more swiftly to repeat presentations of the same antigen.
   c. the ability to distinguish between self and non-self.
   d. the diversity of antigens that we are able to respond to.
   e. all of the above.

Short Answer

5. What is complement?

6. How do cytotoxic T cells kill other cells?

Chapter Test: Answers

1. **T**   2. **T**   3. **e**   4. **b**

5. Complement is a set of proteins that, when activated, can lyse cells.

6. They bind to the cells and release perforins, which are proteins that can lyse the other cells.

Check Your Performance:

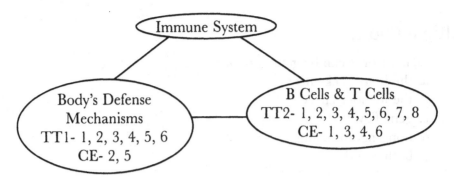

Key: TT = Topic Test; CE = Chapter Exam. Numbers indicate exam questions. Some questions are listed more than once if they refer to more than one topic.

Units III and IV Exam

Multiple Choice

1. Genetic drift can be generated through
   a. the bottleneck effect.
   b. the founder effect.
   c. gene flow.
   d. nonrandom mating.
   e. both a and b.

2. Sexual selection
   a. is the same as nonrandom mating.
   b. refers to the preference of natural selection for one sex, say males, over the other sex (females).
   c. is a form of natural selection.
   d. is required for Hardy-Weinberg equilibrium.
   e. includes all of the above.

3. Comparing a series of spheres of increasing size, the ratio of surface area to volume
   a. increases as size increases, but only very slowly.
   b. increases as size increases, as a function of radius.
   c. stays the same.
   d. decreases as size increases, as a function of 1/radius.

4. Homeostasis
   a. is a form of natural selection.
   b. maintains a relatively constant internal environment in organisms.
   c. is a genetic term referring to the division of chromosomes.
   d. refers to the linkage point for the mitotic spindle apparatus to chromosomes.
   e. relates to humans remaining relatively unchanged during recent evolution.

5. A red blood cell is located in an arteriole in the left leg. How many capillary beds will it pass through before reaching the left ventricle of the heart?
   a. One
   b. Two
   c. Three
   d. Four
   e. None

6. Which of the following hormones is antagonistic to glucagon?
   a. PTH
   b. Insulin
   c. Growth hormone
   d. ACTH
   e. Calcitonin

7. Which of the following hormones is released by the adrenal gland (adrenal medulla) in response to short-term stress?
   a. Epinephrine

b. Mineralocorticoids
c. Glucocorticoids
d. ACTH
e. Cyclic AMP

8. Insulin, a peptide hormone, acts by
    a. binding to the surface of cells and triggering the synthesis of a second messenger inside of the cell.
    b. binding to a receptor, which in turn binds to, or becomes, a transcription factor, enhancing mRNA production.
    c. releasing other hormones from the pancreas.
    d. releasing other hormones from the pituitary.
    e. increasing blood glucose levels.

9. Which of the following is NOT involved in generating or maintaining the resting potential in a neuron?
    a. A high resting potassium conductance
    b. Open potassium channels
    c. The potassium concentration gradient (difference) across the membrane
    d. Voltage-sensitive potassium channels
    e. The sodium-potassium pump

10. Which of the following has a higher concentration inside of a typical neuron than outside?
    a. Sodium
    b. Potassium
    c. Chloride
    d. Calcium

11. An EPSP
    a. is caused by synaptic transmitter binding to receptors.
    b. is a hormone that is released from the adrenal cortex.
    c. is an action potential.
    d. is an electron plus signal peptide.
    e. inhibits action potentials.

12. Which of the following is the first to occur after an action potential arrives at a synapse?
    a. Release of synaptic transmitter
    b. Linkage of synaptic transmitter with receptors
    c. Calcium channels opening
    d. Fusion of synaptic vesicles with the membrane
    e. Degradation of the synaptic transmitter

13. How do antibodies act?
    a. By coating and blocking bacterial toxins
    b. By causing bacteria to clump together
    c. By binding to the surface of foreign cells and activating complement, which in turn destroys the cells
    d. By enhancing phagocytosis
    e. By all of the above ways

14. Consider the signal transduction pathway for a hormone that uses cyclic AMP as a second messenger. Which of the following is correct?

a. Calcium also is involved in this pathway and acts as a third messenger.

b. The cyclic AMP activates diacylglycerol as another signal.

c. The cyclic AMP can activate protein kinases.

d. Adenylyl cyclase makes the cyclic AMP.

e. Both c and d are correct.

15. Which of the following would be consistent with Hardy-Weinberg equilibrium?

a. No change in the frequencies of alleles in the gene pool of a population over several generations

b. No microevolution occurring

c. Random mating

d. Both b and c

e. All of the above

16. Concerning natural selection, which of the following is correct?

a. It is possible for a single allele, in comparison with another allele at the same gene locus, to have both "good" and "bad" effects for an organism.

b. It is not possible for natural selection to shift a population toward two extremes for a trait, while reducing the number of individuals in the middle.

c. Natural selection cannot cause the gene pool for a population to change from one generation to the next.

d. Natural selection cannot occur when genetic drift is occurring.

17. What is a gene pool?

a. A large swimming pool full of Calvin Kleins

b. All of the alleles of all of the genes in an individual

c. All of the alleles of all of the genes in a population

d. All of the alleles of all of the genes in all organisms on the earth

Short Answer

18. Briefly indicate why the following statement is incorrect: Humans have evolved from chimpanzees.

19. A population is in Hardy-Weinberg equilibrium. It has two alleles, A (dominant) and a, at a particular gene locus. One percent of the population is homozygous dominant.

a. At this gene locus, what fraction of the gene pool consists of A alleles?

b. What fraction of the population is heterozygous at this gene locus?

Unit Exam Answers

1. **e**   2. **c**   3. **d**   4. **b**   5. **b**   6. **b**   7. **a**   8. **a**   9. **d**   10. **b**   11. **a**   12. **c**   13. **e**   14. **e**   15. **e**   16. **a**   17. **c**

18. Both humans and chimps have evolved from a common ancestor that existed 5 to 6 million years ago.

19. Since 1% is homozygous dominant, $p^2 = 0.01$. So, $p = 0.1$, or 10% of the alleles in the gene pool are A alleles. The heterozygous percentage must be $2pq = 2 \times 0.1 \times 0.9 = 0.18$, or 18%.

UNIT V:
ECOLOGY

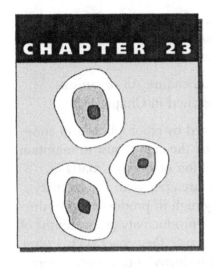

Ecology

Ecology is the study of the interactions between organisms and their environments. Part of the environment of each living organism consists of other living organisms, which are called **biotic** factors. The nonliving, **abiotic** factors include climate, water, and minerals. **Community** refers to all of the living organisms in a given area, and **ecosystem** refers to all of the biotic and abiotic factors in a certain area, thus combining the community with the abiotic components.

There is a web of life. The interactions among organisms and between organisms and environment in an ecosystem are very complex. In this chapter we examine some of the interactions among community members, including food chains and food webs. We look at the cycling of chemicals through ecosystems, and we study the dynamics of population growth.

ESSENTIAL BACKGROUND

- Energy (Chapter 6)
- Photosynthesis (Chapter 8)
- Populations (Chapter 17)

TOPIC 1: FOOD WEBS AND PRODUCTIVITY

KEY POINTS

✓ *What are food chains and webs?*

✓ *What is productivity?*

✓ *What conclusions about food chains can be drawn from a knowledge of energy flow though such chains?*

A **food chain** shows the levels of producers and consumers in a community. The lowest level supports all of the other levels, and is called the **producer** level. Producers synthesize organic compounds, and most, including plants on land and phytoplankton in the waters, do so through photosynthesis. All organisms at higher levels in the chain are **consumers**. The next level above the producers in the food chain comprises the **primary consumers**, which are herbivores that feed directly on the producers. At the level above the primary consumers are the **secondary consumers**, which eat the herbivores, and the food chain continues on to a tertiary or quaternary level.

In reality, food chains are oversimplified. Ecosystems consist more realistically of **food webs**. For example, many organisms can be classified as both primary and secondary consumers: Small rodents might eat both plants and smaller primary consumers, such as insects. Snakes can be both secondary and tertiary consumers, and so on. Such interrelationships create more complex webs, rather than simple chains.

Energy flows in an open-ended way through ecosystems, and up food chains. All organisms require energy for growth, reproduction, and maintenance, as we learned in Chapter 6.

A small fraction of sunlight hitting the surface of the earth is captured by chloroplasts and converted into energy in chemical bonds (Chapter 8). Plants use some of the energy just to maintain themselves. The rest of the energy is used for growth—a net production of organic material. Such **productivity** can be measured as mass produced per surface area per year. Productivity varies in different kinds of ecosystems. Tropical rain forests are very high in productivity, producing over 2 kilograms per square meter each year. Oceans have lower productivity (a few tenths of a kilogram per square meter per year), but are equally important because they cover such a large fraction of the earth's surface (about two-thirds). The net primary productivity of an ecosystem depends on the producers, since that is the level where the energy is first trapped and converted to chemical bonds in organic matter.

Each level in the food chain takes energy from the lower levels, converts some for immediate use, and uses the rest for growth. Only the growth portion can be used by the next level up the chain. The energy available at each level necessarily declines because of inefficiencies. For example, plants use only a fraction of the energy trapped in chloroplasts for growth. Herbivores eat only a fraction of the plants, and are not 100% efficient in energy utilization. They, too, use energy for maintenance as well as growth. There probably is only about a 10% transfer of available energy from one level to the next as we go up the chain.

Notice what this implies about the numbers of organisms produced at each level. There are not very many individuals at the top of a food chain or web, compared to the number generated at the producer level. Many organisms must be produced at the lower levels for each organism at a higher level to consume. Notice also that the loss of biomass at each level up the chain ensures that there cannot be too many levels from the bottom to the top of the chain—the energy runs out.

Topic Test 1: Food Webs and Productivity

True/False

1. Primary consumers consume herbivores.

2. Primary consumers are herbivores.

3. Food chains depict reality better than do food webs.

Multiple Choice

4. Concerning food chains and webs, which of the following is responsible for the net primary productivity of an ecosystem?
   a. Tertiary consumers
   b. Secondary consumers
   c. Primary consumers
   d. Producers

5. In most ecosystems, there are only four or five levels in a food chain or web. The suspected reason for this is that
   a. evolution has not had time to develop more levels.
   b. essential nutrients are lost or reduced from one level to the next. Eventually one or more of them run out.
   c. animals at the highest levels of the food chain are too strong and large for evolution to create any predators of them.
   d. there is a decline in total energy because of energy losses at each level.

Short Answer

6. Do the individual organisms at each level of the food chain have to be larger than those at the next level down the chain?

Topic Test 1: Answers

1. **False.** Primary consumers are herbivores, which consume producers.

2. **True.** Primary consumers consume plants or phytoplankton.

3. **False.** Food webs are more realistic models than food chains because the webs indicate that individual organisms can consume from more than just one lower level in a chain.

4. **d.** It is the producers that harness the energy that results in productivity.

5. **d.** One reaches a level where the energy available at the top of a food chain, in a given area, is insufficient to support organisms at a higher level.

6. While predators tend to be larger than prey, this is not always the case. Predators can attack in packs. Also, at the bottom of the chain, consider trees and the insects that eat them.

TOPIC 2: CHEMICAL CYCLING IN ECOSYSTEMS

KEY POINTS

✓ *What are some of the important nutrients that cycle through ecosystems?*

✓ *What are the cycles of water, carbon, nitrogen, and phosphate?*

To follow the cycling of various nutrients through ecosystems, we need to follow their movements from organisms to inorganic form in soil, water, and atmosphere, to longer-term depositing in rocks, coal, oil, and peat. In most cases, one or more of the chemical elements found in the environment become limiting factors in the growth or reproduction of living organisms. The cycling that these elements undergo can be complex.

One of the simpler cycles is that for **water**, which is essential for life. Water evaporates because of solar energy. It returns from the atmosphere to earth through precipitation. The evaporation-precipitation cycle has an additional complicating factor in that most evaporation is from the ocean, and there is a temporary gain in water on land from precipitation of water that has evaporated from the ocean. Groundwater and runoff from streams and rivers eventually return the water to the ocean, but on its way back some of the water supplies the needs of plants and animals.

A most important chemical cycle is that for **carbon**. Carbon dioxide in the atmosphere is assimilated through photosynthesis in plants and phytoplankton. Some of the carbon that was fixed into organic molecules is respired by the plants, returning carbon dioxide to the atmosphere. Some of the organic material is used for growth and only returned to the atmosphere upon death and decay. Decomposing organisms play a role in the return. Finally, some of the organic material in plants is consumed by animals and used for respiration and growth. That used in cellular respiration is returned directly to the atmosphere. That used for growth is returned when the animals die and decompose, or is used as a food source by other animals.

An additional aspect to carbon cycling sometimes is overlooked. In the oceans, cycling of carbon can involve the interaction of carbon dioxide with water, and with calcium. Bicarbonates and carbonates are formed. The reactions are reversible, forming a carbon dioxide reservoir in the oceans. As a result of these reactions, the oceans contain about 50 times as much carbon as the atmosphere.

Humans have been a net contributor of carbon dioxide to the atmosphere, by burning fossil fuels, such as coal and oil, and burning and clearing forests. The level of carbon dioxide in the atmosphere increased about 12% between 1960 and 1990, and the increase since the beginning of the industrial revolution is estimated to be about 35%. Most scientists now believe that the increased levels of carbon dioxide caused by humans is contributing to **global warming**. Some of the energy in visible sunlight, after striking the earth's surface, is radiated back into the atmosphere as infrared light. Much of this passes through the earth's atmosphere into space. However, carbon dioxide in the atmosphere absorbs this infrared light, resulting in a heating of the atmosphere. The higher level of carbon dioxide is resulting in more trapping, and thus more heating, of the atmosphere. This is the **greenhouse effect**, and carbon dioxide is sometimes referred to as a **greenhouse gas**.

The burning of fossil fuels also contributes to **acid rain** (from oxides of nitrogen and sulfur found in the coal and oil, as mentioned in Chapter 2). The acid rain is disrupting lake and pond ecosystems.

There also is a cycle for **nitrogen**. Nitrogen is the most abundant element in the atmosphere, but most organisms cannot use it in the form of N_2. A few organisms are able to convert nitrogen to ammonia, and other organisms can convert ammonia to nitrites. These can be assimilated by plants. As usual, the plants are the ultimate source for animals. Some nitrogen is released from animals as waste matter. The rest, from both plants and animals, is returned to ammonia when they decompose. Denitrifying bacteria also can return nitrogen to the atmosphere, completing the cycle.

Phosphate undergoes a dual cycle, in different time frames. One occurs over shorter times and involves cycling between organisms and soil as phosphate is taken up by plants (phosphate is not found in the atmosphere); some is transferred to animals who eat the plants; and all eventually is returned to the soil from animal wastes and as both plants and animals decompose. On a much longer, geological time scale, phosphate from runoff can precipitate in bodies of water and in sediment, and eventually form rocks. These rocks, as they are uplifted through geological forces into mountains, for example, can release the phosphate through weathering, returning it to soil.

Industrial extraction of phosphates for use in fertilizers, along with human and industrial wastes, is contributing to increased levels of phosphates and other nutrients in runoff, leading to increased phosphate levels in ponds and lakes, which produces **eutrophication**. Phosphate and nitrogen tend to be limiting nutrients for photosynthesis in most ponds and lakes. As a consequence, eutrophication can result in algal blooms and subsequently in reduced oxygen levels in lakes as a result of decaying matter, and this kills fish and destroys lake ecosystems. More recently, we have been attempting to reduce eutrophication of some of the lakes in the United States, by reducing the dumping of industrial and human wastes into those lakes, in the hopes of recovering former ecosystems.

Topic Test 2: Chemical Cycling in Ecosystems

True/False

1. The phosphate cycle really is two cycles, one short term and the other very long term.

2. Excess phosphate levels in a lake can result in the death of fish.

3. There is a net loss of water to the oceans from land as a result of evaporation of water followed by its precipitation.

Multiple Choice

4. Concerning the carbon cycle,
   a. photosynthesizing organisms are a major net releaser of carbon dioxide to the environment.
   b. the amount of carbon dioxide in the atmosphere has been decreasing with time over the last several decades.
   c. respiration in mitochondria contributes to a return of carbon, in the form of carbon dioxide, to the atmosphere.
   d. all of the above are correct.

5. Concerning global warming,
   a. it appears to be produced by increasing carbon dioxide levels in the atmosphere.
   b. the ultimate source of the energy causing the warming is the heat from burning fossil fuels.
   c. the atmosphere is warming because carbon dioxide absorbs ultraviolet light.
   d. all of the above are correct.

Topic Test 2: Answers

1. **True.** The short-term one cycles between living organisms and soil. The long-term cycle involves rock formation.

2. **True.** This occurs through eutrophication.

3. **False.** The opposite is true. There is a net gain of water from precipitation that has come from ocean evaporation.

4. **c.** Cellular respiration releases carbon dioxide, and this occurs in mitochondria, as described in Chapter 7.

5. **a.** The ultimate energy source is sunlight, not heat from burning the fossil fuels. Infrared light, not ultraviolet light, is being absorbed by the carbon dioxide to generate the warming.

TOPIC 3: LIFE HISTORY CHARACTERISTICS AND POPULATION DYNAMICS

KEY POINTS

✓ *What are the two major extremes in life history characteristics?*

✓ *What are two important models of population growth?*

Species have evolved different **life history** characteristics, and these influence how populations change in number. The life histories of organisms can be viewed as resulting, in part, from evolutionary "decisions" about the use of energy. Most organisms appear to be limited by their sources of energy. The resulting energy "budget" can be "spent" in different ways. We consider two extremes. At one extreme are **r-selected** or **opportunistic** species, and at the other are **K-selected** or equilibrial species. The major differences are shown in **Table 23.1**.

Each set of properties has its advantages when it comes to natural selection. It is perhaps more obvious that rapid maturation and the production of many offspring carry a selective advantage, especially for species that have high mortality (death) rates. Darwinian fitness is measured by the contribution an organism makes to the gene pool. The downside for opportunistic (r-selected) organisms is that the production of large numbers of offspring means less investment can be made in each one. Offspring of opportunistic species usually receive little care or protection from parents, and most die before reaching reproductive age. Each offspring has less energy invested in it, and experiences a greater risk of death. At the same time, there are many more offspring of r-selected species than there are of K-selected organisms, so only a few have to survive, and in favorable environments, r-selected organisms are capable of explosive growths in numbers.

The distinction between opportunistic and equilibrial species is an oversimplification. Some species do not fit well; other species fit in some ways, but not others. Consider oak trees, which are K-selected in many ways, but produce many small acorns, with little energy invested in each. Climate, predation, and stress influence the evolution of life history traits in complex ways. Nevertheless, the distinction between opportunistic and equilibrial species can be useful in understanding some aspects of population ecology.

Two simplified models of population growth are used often. Ignoring migrations, only two factors influence population size over time, and these are the numbers of births and deaths. If a population is growing in proportion to its size, then $\frac{\Delta N}{\Delta t} = rN$, where N represents the number

Table 23.1 The Major Differences Between r-selected or Opportunistic Species, and K-selected or Equilibrial Species

PROPERTY	r-SELECTED	K-SELECTED
Maturation	Rapid	Slow
No. of offspring	Many	Few
Bouts of reproduction	One	Repeated
Investment in homeostasis and maintenance	Less	Greater
Initial mortality rates	High	Lower
Life spans	Short	Longer
Example organism	Insects, dandelions	Humans, elephants

of individuals in the population and t represents time. The population growth rate, r, is the birth rate minus the death rate. If r is positive, the population is growing; if r is negative, the population is shrinking; and an r of zero indicates a stable population size. We speak of r_{max} as the fastest rate of increase possible for a population.

With r higher than zero, this model gives exponential growth. An equivalent equation then is $N = N_0e^{rt}$, where N_0 is the number in the population when t is zero. It is useful to realize that exponential growth gives a constant doubling time. If it takes 40 years for the population to double from 1,000 to 2,000 individuals, it will take another 40 years for it to reach 4,000, and the population will be at 8,000 forty years after that. This is a simplified view of population growth and describes what can happen to a population under ideal conditions, with unlimited space and resources. Such conditions can be generated in the laboratory, and can be found under certain conditions, for a few population doublings, in nature.

However, life and population growth are not usually so simple. Simple exponential growth cannot continue for very many doublings before something gives out. Food might become limiting, for example. A second model takes into account such limits. It is called **logistic growth**, and presumes a limit on the number of organisms within a population that can be sustained by an ecosystem. By adding a term to the exponential growth equation, we can model logistic growth: $\frac{\Delta N}{\Delta t} = rN[(K - N)/K]$. The constant, K, is called the carrying capacity for the population. We quickly get a sense of what the addition of the new term does. If N is much less than K, that is, if the population is small, the ratio $[(K - N)/K]$ has a value close to one, and the result is little different from exponential growth. So, when population size, N, is small relative to K, growth is roughly exponential. As N approaches K, the ratio $[(K - N)/K]$ approaches zero, and so as the population size approaches K, the growth will go to zero. This is logistic growth—exponential growth at first, but slowing as a plateau value is reached. K is that plateau value; it is the maximum number of individuals that the environment can sustain, that is, the maximum stable population size for that environment.

We now see why opportunistic species are called r-selected, since natural selection appears to have maximized the population growth rate potential. Equilibrial species are K-selected because they tend to have populations that stabilize near their carrying capacity, K. In nature, r-selected organisms typically undergo large fluctuations in numbers. They might die off in winter and rapidly multiply each summer, for example.

Both of these population models are vastly oversimplified. Using such models in nature usually is much more complicated. Some populations may continue to oscillate in number, depending on food availability and predators. Some overshoot a K value, and then dramatically die off as resources in the environment are depleted. Many factors influence population sizes and growth rates.

Topic Test 3: Life History Characteristics and Population Dynamics

True/False

1. Opportunistic organisms invest more in homeostasis and maintenance than do equilibrial organisms.

2. Logistic growth is more rapid than exponential growth.

Multiple Choice

3. Which of the following is characteristic of an r-selected (opportunistic) organism?
   a. Slow maturation
   b. Repeated bouts of reproduction
   c. Having many offspring
   d. Long life span
   e. Low initial mortality rates

4. A population is exhibiting exponential growth. It doubles in size from the year 1950 to 1990. When will it be four times the 1950 population level?
   a. 2010
   b. 2030
   c. 2050
   d. 2070
   e. 2090

5. For logistic growth of the form $\frac{\Delta N}{\Delta t} = rN[(K - N)/K]$, which of the following correctly describes what happens when N is much less than K?
   a. N grows at nearly an exponential rate.
   b. The growth rate for N is near to zero.
   c. The value for r goes to zero.
   d. The value for K goes to zero.
   e. N only approaches K at time t = 0.

Short Answer

6. A population is growing exponentially. If it takes 1 year for a population to double, how long does it take for the population to grow to 16 times its original size?

Topic Test 3: Answers

1. **False.** Equilibrial organisms invest more.

2. **False.** Exponential growth is more rapid. Logistic growth is like damped exponential growth that reaches a plateau.

3. **c.** All of the other choices are characteristics of equilibrial organisms.

4. **b.** It takes the population 40 years to double, so it will double again between 1990 and 2030.

5. **a.** With N small relative to K, the equation simplifies to approximately the exponential growth equation, $\frac{\Delta N}{\Delta t} = rN$.

6. **Four years.** Doubling rate is a constant for an exponentially growing population. It takes 1 year to double, so in 2 years the population will double again, to give four times its initial size. A third year results in a population eight times the initial size, and 4 years results in a population 16 times the original size.

APPLICATION: HUMAN POPULATION GROWTH

We can, with some estimates, trace the total number of humans on earth during the last several thousand years. Until about 500 years ago, our numbers were growing very slowly, with some ups and downs caused by plagues and famine. With the industrial revolution and improvements in agriculture and sanitation, human death rates have declined dramatically. For the last several hundred years, human population growth has been looking more like that of an r-selected species than a K-selected species. What has actually happened is an increase in the short-term carrying capacity, K, through technology. We have not only been growing exponentially, but, until recently, at an even greater than exponential rate. Between 1650 and 1850, the human population on earth doubled from $\frac{1}{2}$ to 1 billion. Based on exponential growth, that number should double again, to reach 2 billion in 2050, but a population of 2 billion was reached in only 80 years, by 1930. By 1975 the human population had reached 4 billion, doubling again in just 45 years. As the end of the century approaches, the number of humans on earth is close to 6 billion.

The dramatic growth in human population has produced a number of impacts on the biosphere. Ecosystems have been disrupted by human activities—some intentionally, some unintentionally. Earlier in this chapter we discussed global warming, acid rain, deforestation, and eutrophication. We also have a growing amount of waste, in various forms. In south Florida solid wastes have already created what is called Mt. Trashmore on an otherwise flat landscape.

Other human-related alterations include **ozone layer depletion** caused by human-produced chemicals such as chlorofluorocarbons. Depletion of the ozone layer has increased the amount of ultraviolet light reaching the surface of the earth. The ultraviolet light contributes to skin cancers and to eye problems. Fortunately, international agreements have reduced the production of ozone-depleting chemicals, but the damage will continue for decades because of the long lifetime of some of the chemicals in the atmosphere.

Humans also have introduced **exotic species**, often accidentally, into environments. In south Florida, two exotic plants, Brazilian pepper and melaleuca trees, are dramatically altering the landscape, crowding out native species.

Some human-introduced chemicals have proved to be damaging. One of these, **DDT**, an insecticide, caused birds of prey such as bald eagles and osprey to decline greatly in number. Hydrophobic DDT tended to concentrate in the membranes and fat tissues of organisms, as described in Chapter 4. Thus, DDT reached higher and higher concentrations at each higher level of the food chain, from phytoplankton at the producer end to the fish that the birds ate. Elimination of DDT in the United States has resulted in a resurgence in the number of these majestic birds.

An increasingly serious result of human activities has been the **extinction of other species**. We destroy existing ecosystems as we create farms to make food for human consumption, and such ecosystem destruction results in unfavorable environments for many species. For example, farming near the equator is causing the destruction of rain forests, which carries with it a considerable loss of **biodiversity**, a measure of the number of species and relative abundance of each different species. Throughout the world, we are in the middle of a human-caused extinction of species the likes of which has not been seen since the loss of dinosaurs and other species, apparently caused by a large meteor hitting earth 65 million years ago.

Humans also are using the earth's resources, such as coal and oil, at nonsustainable rates. It is not at all clear how much longer our population can continue to grow without dire consequences. Developed countries tend to have lower growth rates, because their birth rates have dropped, but these same developed countries also tend to have higher per-person utilization of resources. Today we see famines in parts of Africa, and the AIDS epidemic, with millions infected worldwide, may ultimately match the number of deaths caused by the plague. We can hope that these are not hints of worse to come.

One thing should be obvious: We cannot keep growing forever. It may be difficult to predict exactly what the practical limit in the number of humans will be for earth, but a limit there must be. In the meantime, as numbers grow, the quality of life can decline. One only has to consider the traffic on the roads in the United States to realize that numbers can limit quality. It is not clear how the end of human population growth will come—by a logistic-like, gradual decline in growth rate or by an overshoot in numbers followed by a dramatic loss of human life from famine, war, or other catastrophe. Unlike other species, humans have the knowledge and ability to control birth rates, and so avoid an environmental or human catastrophe. It remains to be seen whether enough of us have the will to do so.

Chapter Test

True/False

1. A community consists of all of the living organisms in a given area.

2. Productivity refers to the amount of organic matter generated per area per year.

3. Humans are using resources at a sustainable rate.

Multiple Choice

4. What contributed to DDT concentrating in biological tissues?
   a. It was concentrated in membranes.
   b. It was secreted from animals in urine.
   c. It was concentrated in the bark of trees.
   d. It was absorbed from nest materials into birds' egg shells.

5. Which of the following is NOT a life history characteristic of r-selected (opportunistic) organisms?
   a. Slow maturation
   b. Many offspring
   c. One bout of reproduction
   d. Smaller investment in maintenance
   e. High initial mortality

6. Primary consumers
   a. produce chemical energy from sunlight.
   b. consume secondary consumers.

c. are herbivores.

d. are at the bottom of the food chain.

e. are at the top of the food chain.

Short Answer

7. What is the major constituent of automobile exhaust that contributes to greenhouse gases and might cause global warming?

8. Is ozone layer depletion the major cause of global warming?

Chapter Test: Answers

1. **T** 2. **T** 3. **F** 4. **a** 5. **a** 6. **c**

7. Carbon dioxide is the major constituent. Carbon dioxide and water are the two major, unavoidable products of the combustion of gasoline.

8. No, ozone layer depletion contributes to the amount of ultraviolet light reaching earth's surface, but this is a minor contributor to global warming. The greenhouse effect is more major. It results from infrared light, irradiated from the earth's surface in response to sunlight hitting the earth. The infrared light is absorbed by carbon dioxide in the atmosphere, resulting in a warming of the air. The greater the concentration of carbon dioxide in the atmosphere, the greater the amount of infrared light absorbed, and thus, the greater the warming of the lower atmosphere.

Check Your Performance:

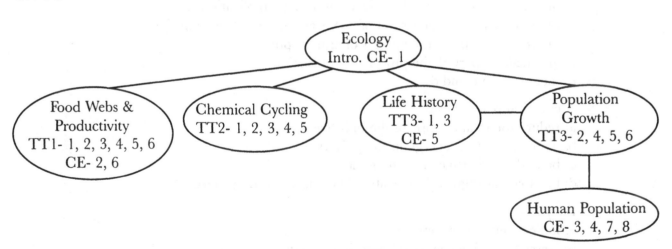

Key: TT = Topic Test; CE = Chapter Exam. Numbers indicate exam questions. Some questions are listed more than once if they refer to more than one topic.

Final Exam

Multiple Choice

1. The most abundant elements in living organisms are those commonly found in amino acids, lipids, and nucleic acids. These are
   a. C, H, N, O, P, S.
   b. C, H, O, P, Fe.
   c. C, H, O, Na, K, Cl.
   d. C, H, O, Na, Ca, K.
   e. C, H, O, Na, Fe, K.

2. Tertiary structure in proteins is held in place by which of the following forces?
   a. Hydrogen bonds
   b. Disulfide bonds
   c. Ionic bonds
   d. Hydrophobic interactions
   e. All of the above

3. A piece of DNA in solution is directly taken up by a bacterial cell and is incorporated into the bacterium's genetic material. This is called
   a. transduction.
   b. transformation.
   c. conjugation.
   d. maternal inheritance.

4. Okazaki fragments
   a. are found on both strands of DNA at a single replication site.
   b. are found on only one strand of DNA at a single replication site.
   c. are made as part of the process of transcription.
   d. are made as part of the process of translation.
   e. include both c and d.

5. A chemical reaction can
   a. reduce the total entropy in the universe.
   b. increase the total energy in the universe.
   c. be both exothermic and exergonic.
   d. be at equilibrium with a positive ΔG (change in free energy).

6. Enzymes
   a. increase energy of activation.
   b. increase the free energy released in a reaction.
   c. decrease energy of activation.
   d. decrease the free energy released in a reaction.
   e. do both a and b.

7. What kind of junctions between skin cells are important for blocking the passage of substances through the skin?
   a. Tight junctions
   b. Gap junctions

c. Desmosomes

d. Synaptic junctions

8. One strand on a DNA molecule has the sequence GATTACA. The other DNA strand has the sequence

a. GATTACA.

b. CTAATGT.

c. AGCCGTG.

d. TCGGCAC.

e. AGUUGCG.

9. Concerning the carbon cycle,

a. assimilation of carbon dioxide from the atmosphere occurs primarily in chloroplasts.

b. generation of the carbon dioxide that passes from living organisms to the atmosphere occurs primarily in mitochondria.

c. more carbon dioxide is found in the oceans than in the atmosphere.

d. both b and c are correct.

e. all of the above are correct.

10. K-selected (equilibrial) organisms tend to have all of the following characteristics EXCEPT:

a. slow maturation.

b. repeated bouts of reproduction.

c. lower initial mortality rates.

d. many offspring in each reproductive bout.

e. longer life spans.

11. Which of the following is NOT an example of active transport across membranes?

a. The action of the Na pump (Na-K ATPase)

b. Co-transport

c. Facilitated diffusion

d. Both b and c

e. All of the above

12. Microevolution

a. is occurring whenever the frequency of alleles in a gene pool is changing.

b. has natural selection as its only cause.

c. occurs whenever there is Hardy-Weinberg equilibrium.

d. is the study of the evolution of micros.

e. includes both b and c.

For the next three questions, consider the following cross: Aabb × aaBb, where A and B are dominant alleles, and a and b are recessive.

13. In the above cross, how many different gametes can result from the individual on the right?

a. One

b. Two

c. Three

d. Four

e. None

14. How many different genotypes will there be among the first-generation offspring from the above cross?
    a. One
    b. Two
    c. Three
    d. Four
    e. None of the above

15. In the above cross, how many different phenotypes will there be among the first-generation offspring?
    a. One
    b. Two
    c. Three
    d. Four
    e. None of the above

16. A male with a sex-linked recessive trait marries a female who is homozygous dominant at that gene locus. What fraction of their children are expected to have the disorder?
    a. None
    b. 1.0
    c. 0.5
    d. 0.25
    e. Half of the males, none of the females

17. A cell whose diploid DNA content is 6 picograms is in the G_2 phase of the cell cycle. The amount of DNA in the cell will be
    a. 3 picograms.
    b. 6 picograms.
    c. somewhere between 6 and 12 picograms.
    d. 12 picograms.
    e. 24 picograms.

18. During photosynthesis, oxygen is formed
    a. during cyclic electron flow in the light reactions.
    b. during noncyclic electron flow in the light reactions.
    c. during the Calvin cycle.
    d. at the end of the electron transport chain.
    e. by ATP synthase.

19. During cell respiration, carbon dioxide is released
    a. as pyruvate is converted to acetyl-CoA.
    b. during the Krebs cycle.
    c. during glycolysis.
    d. during electron transport.
    e. during both a and b.

20. A molecule that donates electrons to the electron transport chain in mitochondria is
    a. ATP.
    b. water.
    c. NADH.
    d. NAD+.
    e. carbon dioxide.

21. Comparing mitosis and meiosis, which of the following is correct?
    a. Both are involved in the process of distributing genetic material to daughter cells.
    b. Both involve crossing-over.
    c. Both result in two daughter cells.
    d. Both distribute identical genetic material to daughter cells.
    e. Both a and d are correct.

22. IPSPs
    a. are a type of action potential.
    b. can generate action potentials.
    c. are synaptic potentials.
    d. are both b and c.
    e. are all of the above.

23. Which of the following ions directly triggers the release of synaptic transmitters from presynaptic nerve terminals?
    a. Sodium
    b. Potassium
    c. Chloride
    d. Magnesium
    e. Calcium

24. What does parathyroid hormone (PTH) do?
    a. Increases metabolism in most body cells
    b. Decreases metabolism in most body cells
    c. Raises blood calcium levels
    d. Raises blood glucose levels
    e. Lowers blood glucose levels

25. Which of the following cells most directly gives our immune system the ability to produce antibodies to a second exposure to an antigen (secondary immune response)?
    a. B cells
    b. Memory B cells
    c. T cells
    d. Macrophages

26. After an action potential is initiated, why does it propagate?
    a. Because potassium ions flow out of the cell at the end of the action potential
    b. Because the sodium ion channels inactivate
    c. Because the inside of the cell becomes negatively charged
    d. Because the depolarization of the cell that occurs during an action potential spreads inside the axon
    e. Because of a negative feedback loop involving sodium ions

27. A blood cell is in a capillary in the right lung. What chamber of the heart will it return to first?
    a. Right atrium
    b. Right ventricle
    c. Left atrium
    d. Left ventricle

28. Blood pressure is lowest
    a. in arteries.
    b. in capillaries.
    c. in veins.
    d. where the velocity of blood flow is the lowest.
    e. in both c and d.

29. Restriction enzymes
    a. can be used to make RFLPs.
    b. are made by bacteria.
    c. cut DNA at particular sequences.
    d. include both b and c.
    e. include all of the above.

30. Lysogenic viruses
    a. can undergo a lytic cycle of virus production in a bacterial cell.
    b. can incorporate their DNA into the DNA of the host.
    c. can remain dormant for a number of generations inside the host cell, getting copied each time the host DNA is.
    d. include all of the above.
    e. include none of the above.

31. Charles Darwin
    a. was the first to propose that evolution occurs.
    b. was the first to put together substantial evidence in favor of evolution.
    c. put forward the idea of natural selection.
    d. did both b and c.
    e. did all of the above.

32. Which of the following would necessarily violate the second law of thermodynamics without necessarily violating the first law?
    a. Matter is created out of nothing.
    b. Chemical energy increases while electrical energy decreases, and a little heat is generated.
    c. A container is initially at uniform temperature throughout, but one-half becomes hotter and the other cooler, with no other change occurring.
    d. A rock rolls down a hill.
    e. All of the above satisfy the requirements.

33. In the light reactions of photosynthesis,
    a. primarily green light is absorbed by chloroplasts.
    b. ATP is formed.
    c. NADPH is formed.
    d. carbon is fixed into organic compounds.
    e. both b and c are correct.

34. Comparing Krebs and Calvin cycles,
    a. both involve the generation of carbon dioxide.
    b. both involve the net production of organic molecules from carbon dioxide.
    c. both generate ATP.
    d. both initially begin with acetyl groups from acetyl-CoA.
    e. none of the above are correct.

35. Transcription in eukaryotic cells
    a. requires RNA polymerase enzymes.
    b. can produce RNA molecules that contain introns and exons.
    c. can involve enhancer sequences.
    d. can involve transcription factors.
    e. involves all of the above.

36. A molecule with the molecular formula $C_3H_7O_2N$ is most likely to be
    a. a lipid.
    b. a sugar.
    c. an amino acid.
    d. a steroid hormone.
    e. DNA.

37. A membrane separates two solutions. Solution 1 contains 0.5 molar NaCl, and solution 2 contains 1 molar NaCl. The membrane is permeable only to Na^+, not Cl^-. After time has elapsed,
    a. both solutions will contain the same amount of sodium, but there will be less chloride in solution 1.
    b. a little net movement of sodium from solution 2 to solution 1 will have occurred.
    c. solution 1 will have a positive charge relative to solution 2.
    d. both a and c are correct.
    e. both b and c are correct.

38. A major role of the Golgi apparatus is
    a. to sort and package membrane-bound materials.
    b. to synthesize membrane proteins and lipids.
    c. to make ribosomes.
    d. to digest macromolecules.
    e. none of the above.

39. Chloroplasts and mitochondria
    a. probably arose from prokaryotic cells that invaded eukaryotic cells.
    b. contain DNA.
    c. are found in plant cells.
    d. All of the above are correct.

40. How many ATP molecules are made from each glucose molecule during fermentation?
    a. One
    b. Two
    c. Three
    d. Four
    e. About 36

Final Exam: Answers

1. **a**   2. **e**   3. **b**   4. **b**   5. **c**   6. **c**   7. **a**   8. **b**   9. **e**   10. **d**   11. **c**   12. **a**   13. **b**
14. **d**   15. **d**   16. **a**   17. **d**   18. **b**   19. **e**   20. **c**   21. **a**   22. **c**   23. **e**   24. **c**   25. **b**
26. **d**   27. **c**   28. **c**   29. **e**   30. **d**   31. **d**   32. **c**   33. **e**   34. **e**   35. **e**   36. **c**   37. **e**
38. **a**   39. **d**   40. **b**

INDEX